Rust for Blockchain Application Development

Learn to build decentralized applications on popular blockchain technologies using Rust

Akhil Sharma

Rust for Blockchain Application Development

Group Product Manager: Kaustubh Manglurkar

Publishing Product Manager: Arindam Majumder

Book Project Manager: Hemangi Lotlikar

Senior Editor: Vandita Grover

Technical Editor: Kavyashree K S

Copy Editor: Safis Editing

Proofreader: Safis Editing

Indexer: Rekha Nair

Production Designer: Shankar Kalbhor

Senior DevRel Marketing Executive: Nivedita Singh

First published: April 2024

Production reference: 1290324

Published by

Packt Publishing Ltd.

Grosvenor House

11 St Paul's Square

Birmingham

B3 1RB, UK.

ISBN 978-1-83763-464-4

www.packtpub.com

To the women in my life – my mother, Manisha Sharma, for all her sacrifices; my sister, Neha Sharma, for being a constant source of encouragement; and my wife, Akanksha, for being extremely supportive throughout this journey.

– Akhil Sharma

Contributors

About the author

Akhil Sharma is the founder of Armur AI, a cybersecurity company that is backed by Techstars, Outlier Ventures, and Aptos, and is part of the Google AI startups cloud program.

Akhil teaches advanced engineering topics (Rust, Go, Blockchain, and AI) on his YouTube channel and has mentored more than 200,000 engineers across platforms such as Linkedin Learning, Udemy, and Packt.

Being deeply involved with multiple Rust-based blockchain communities such as Aptos, Solana, and Polkadot inspired Akhil to write this book.

In his free time, Akhil likes to train in jiu jitsu, play the guitar, and surf.

About the reviewers

Denis Cavalli is a lead software engineer with a strong background in embedded systems, software development, and R&D. He graduated in computer engineering from the Universidade Federal do Amazonas in Brazil, and has more than 10 years of experience in software development and team leadership, working for start-ups and big companies.

Since 2021, he has been engaged with the Web3 environment, experimented with Ethereum/Solidity and Solana, worked professionally for Web3 companies using the Helium SDK, designed decentralized solutions targeted for Polkadot/Kusama networks using Substrate, and has had smart contracts deployed on the Arbitrum Nova mainnet.

Ryu Kent is a senior blockchain engineer who has worked in the industry for 7 years. He is particularly active in the DAO space and has launched a number of well-known smart contracts. Prior to moving to Web3, Ryu spent over a decade working in financial services, including HSBC, Barclays Bank, and PriceWaterhouseCoopers, building centralized ledgers.

Table of Contents

2

Part 2: Building the Blockchain

3

4

Adding More Features to Our Custom Blockchain   97

5

Finishing Up Our Custom Blockchain   127

Part 3: Building Apps

6

7

8

Exploring NEAR by Building a dApp   243

Part 4: Polkadot and Substrate

9

10

Part 5: The Future of Blockchains

11

Preface

Rust is one of the most widely used languages in blockchain systems and many popular blockchains including Solana, Polkadot, Aptos, and Sui are built with Rust. Rust frameworks such as Foundry are also highly preferred by developers of established chains including Ethereum.

Learning how decentralized apps work on popular Rust chains and also how to build your own blockchains – whether from scratch or using frameworks such as Substrate – is an important skill to have since all big dApps, at some point, end up moving to their own chains, also referred to as *application chains*.

This book is for developers who want to go deep and understand how Rust is used for building dApps and blockchains and add a new dimension to their Rust skills.

Who this book is for

This book is for blockchain and dApp developers, blockchain enthusiasts, and Rust engineers who want to step up their game by adding blockchain to their repertoire of skills.

What this book covers

Chapter 1, Blockchains with Rust, outlines the critical blockchain concepts that we will use in the book.

Chapter 2, Rust – Necessary Concepts for Building Blockchains, explores the critical Rust concepts that we will be using to build our own blockchain.

Chapter 3, Building a Custom Blockchain, lays the foundation and the building blocks for our own custom blockchain that we're building from scratch.

Chapter 4, Adding More Features to Our Custom Blockchain, sees up build on our blockchain and add more features to it.

Chapter 5, Finishing Up Our Custom Blockchain, brings together all the individual blocks that we have built and combines them into a complete blockchain.

Chapter 6, Using Foundry to Build on Ethereum, explores Foundry, a Rust framework that can be used to build and deploy smart contracts on Ethereum.

Chapter 7, Exploring Solana by Building a dApp, teaches you how to build a dApp for Solana.

Chapter 8, Exploring NEAR by Building a dApp, teaches you how to build a dApp for an upcoming blockchain, NEAR.

Chapter 9, Exploring Polkadot, Kusama, and Substrate, explores the basic concepts behind Substrate, which enables developers to build their own chains.

Chapter 10, Hands-On with Substrate, uses our knowledge of Substrate to build a custom blockchain.

Chapter 11, Future of Rust for Blockchains, discusses the future of blockchains with Rust.

To get the most out of this book

We're assuming that you know your way around Rust and have knowledge of all its basic concepts.

Software/hardware covered in the book	Operating system requirements
Rust 1.74.0 or higher	Windows, macOS, or Linux
Cargo	Windows, macOS, or Linux

If you are using the digital version of this book, we advise you to type the code yourself or access the code from the book's GitHub repository (a link is available in the next section). Doing so will help you avoid any potential errors related to the copying and pasting of code.

Download the example code files

You can download the example code files for this book from GitHub at `https://github.com/PacktPublishing/Rust-for-Blockchain-Application-Development`. If there's an update to the code, it will be updated in the GitHub repository.

We also have other code bundles from our rich catalog of books and videos available at `https://github.com/PacktPublishing/`. Check them out!

Conventions used

There are a number of text conventions used throughout this book.

`Code in text`: Indicates code words in text, database table names, folder names, filenames, file extensions, pathnames, dummy URLs, user input, and Twitter handles. Here is an example: "rustup is the toolchain manager that includes the compiler and Cargo's package manager."

A block of code is set as follows:

```
pub struct Block {
    timestamp: i64,
    pre_block_hash: String,
    hash: String,
    transactions: Vec<Transaction>,
    nonce: i64,
    height: usize,
}
```

Any command-line input or output is written as follows:

```
brew install rustup
```

Bold: Indicates a new term, an important word, or words that you see onscreen. For instance, words in menus or dialog boxes appear in **bold**. Here is an example: "Working with strings is straightforward in Rust, so it's important to know the difference between the **String type** and **string literals**."

> **Tips or important notes**
> Appear like this.

Get in touch

Feedback from our readers is always welcome.

General feedback: If you have questions about any aspect of this book, email us at customercare@ packtpub.com and mention the book title in the subject of your message.

Errata: Although we have taken every care to ensure the accuracy of our content, mistakes do happen. If you have found a mistake in this book, we would be grateful if you would report this to us. Please visit www.packtpub.com/support/errata and fill in the form.

Piracy: If you come across any illegal copies of our works in any form on the internet, we would be grateful if you would provide us with the location address or website name. Please contact us at copyright@packt.com with a link to the material.

If you are interested in becoming an author: If there is a topic that you have expertise in and you are interested in either writing or contributing to a book, please visit authors.packtpub.com.

Share Your Thoughts

Once you've read *Rust for Blockchain Application Development*, we'd love to hear your thoughts! Scan the QR code below to go straight to the Amazon review page for this book and share your feedback.

https://packt.link/r/1837634645

Your review is important to us and the tech community and will help us make sure we're delivering excellent quality content.

Download a free PDF copy of this book

Thanks for purchasing this book!

Do you like to read on the go but are unable to carry your print books everywhere?

Is your eBook purchase not compatible with the device of your choice?

Don't worry, now with every Packt book you get a DRM-free PDF version of that book at no cost.

Read anywhere, any place, on any device. Search, copy, and paste code from your favorite technical books directly into your application.

The perks don't stop there, you can get exclusive access to discounts, newsletters, and great free content in your inbox daily

Follow these simple steps to get the benefits:

1. Scan the QR code or visit the link below

https://packt.link/free-ebook/9781837634644

2. Submit your proof of purchase

3. That's it! We'll send your free PDF and other benefits to your email directly

Part 1: Blockchains and Rust

In this part, we will first get some knowledge about blockchains and some necessary Rust concepts that'll help us in building a fully fledged blockchain.

This part has the following chapters:

- *Chapter 1, Blockchains with Rust*
- *Chapter 2, Rust - Necessary Concepts for Building Blockchains*

1

Blockchains with Rust

Blockchains have a lot of mystery around them, and only a few engineers have complete clarity of the inner workings and how disruptive they will be to the incumbent way of working for many industries.

With the help of this chapter, we want to tackle the very core concepts of blockchains. Since this is a book about using Rust for blockchains, we want to, at the same time, understand why Rust and blockchains are a match made in heaven. This will also provide us with insight into why some popular blockchains (Solana, Polkadot, and NEAR) have used Rust and why the latest blockchains to enter the market (Aptos and Sui) are also choosing Rust above any other technology that exists on the market today.

The end goal of this chapter is to provide a comprehensive understanding of the critical concepts around blockchains that will enable us to build a blockchain from scratch later in the book.

In this chapter, we're going to cover the following main topics:

- Laying the foundation with the building blocks of blockchains
- Exploring the backbone of blockchains
- Understanding decentralization
- Scaling the blockchain
- Introducing smart contracts
- The future of the adoption of blockchains

Laying the foundation with the building blocks of blockchains

In this section, let's learn the most basic concept of blockchains—what a blockchain is made up of.

A **blockchain** can be imagined as a series of connected blocks, with each **block** containing a finite amount of information.

The following diagram demonstrates this clearly with multiple connected blocks.

Figure 1.1 – Representation of a blockchain

Just like in a traditional database, there are multiple tables in which the data is stored sequentially in the form of **records**, and the blockchain has multiple blocks that store a particular number of **transactions**.

The following diagram demonstrates blocks as a store for multiple transactions:

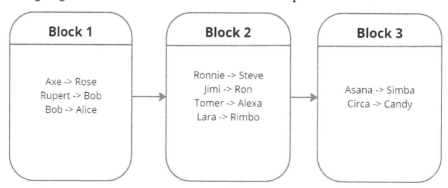

Figure 1.2 – Blocks with transaction data

The question now is, why not just use databases? Why do we even need blockchains? Well, the main difference here is that there is no admin and nobody is in charge. The other significant difference is that most blockchains are engineered to be **permissionless** at the core (even though **permissioned** blockchains exist and have specific use cases at the enterprise level), making them accessible to everyone and not just to people with access.

Another equally substantial difference is that blockchains only have insert operations, whereas databases have **CRUD** operations, making blockchains inherently **immutable**. This also implies that blockchains are not **recursive** in nature; you cannot go back to repeat a task on records while databases are recursive.

Now, this is a complete shift in how we approach data storage with blockchains in comparison to traditional databases. Then there is **decentralization**, which we will learn about shortly and that is what makes blockchains an extremely powerful tool.

Web 3.0, another confusing and mysterious term, can, at a considerably basic level, be defined as the internet of blockchains. Until now, we have had **client-server architecture** applications being connected to each other. That was Web 2.0, but suddenly, with the help of blockchains, we will have a more decentralized internet. Even if most of this does not make sense right now, do not despair, for we have plenty to cover.

In the following subsections, we will learn about things such as hashes, transactions, security, decentralized storage, and computing.

Blocks

The smallest or atomic part of any blockchain is a **block**. We learned in the previous section that blocks contain transactions, but that's not all; they also store some more information. Let's peel through the layers.

Let's look at a visual representation of the inner workings of a block:

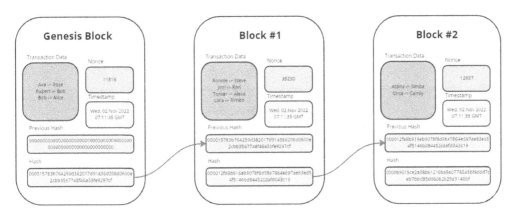

Figure 1.3 – Connected blocks of a blockchain

In the preceding diagram, we notice that the first block is called the **Genesis Block**, which is an industry-standard term for the first block of the chain. Now, apart from transaction data, you also see a **hash**. In the next section, *Hashes*, we will learn how this hash is created and why it is required. For now, let's consider it to be a random number. So, each block has a hash, and you will also notice that the blocks are storing the *previous hash*. This is the same as the hash of the previous block.

The *previous hash* block is critical because it is what *connects* the blocks to each other. There is no other aspect that connects the blocks to make a blockchain; it's simply the fact that a subsequent, sequential block holds the hash of the previous block.

We also notice a field called **nonce**. This stands for **number only used once**. For now, we need to understand that the nonce needs to be consistent with the hash for the block to be valid. If they're not consistent, the following blocks of the blockchain go completely out of sync and this fortifies the *immutability* aspect of blockchains that we will learn about in detail in the *Forking* section. Now, as we go further, we will uncover more layers to this, but we're at a great starting point and have a broad overview.

Hashes

Hashes are a core feature of the blockchain and are what hold the blocks together. We remember from earlier that blocks store hash and previous hash and hashes are simply created by adding up all the data, such as transactions and timestamps, and passing it through some hashing algorithm. One example is the **SHA-256** algorithm.

The following diagram shows a visual representation of data being passed to the SHA-256 algorithm and being converted into a usable hash:

Figure 1.4 – Data to SHA-256 hash

A hash is a unique fixed-length string that can be used to identify or represent a piece of data and a hash algorithm, such as SHA-256, is a function that computes data into a unique hash.

While there are several other SHA algorithms available (such as **SHA-512**), SHA-256 stands as the most prevalent choice within blockchains due to its robust hash security features and the notable fact that it remains unbroken to this day.

There are four important properties of the SHA-256 algorithm:

- One-way: The hash generated from SHA-256 is 256 bits (or 32 bytes) in length and is irreversible; if you want to get the plaintext back (plaintext being the data that we passed through SHA-256), you will not be able to do so.

- Deterministic: Every time you send a particular data through the algorithm, you will get the same predictable result. This means that the hash doesn't change for the same data.

- Avalanche effect: Changing one character of the data, completely changes the hash and makes it unrecognizable.

- For example, the hash for *abcd* is

  `88d4266fd4e6338d13b845fcf289579d209c897823b9217da3e161936f031589`

  but the hash for *abce* is

  `84e73dc50f2be9000ab2a87f8026c1f45e1fec954af502e9904031645b190d4f.`

- The only thing common between them is that they start with 8. There's nothing else that matches, so you can't possibly predict how the algorithm represents *a*, *b*, or *c*, and you can't work your way backward to either get the plaintext data or predict what the hash representation for some other data will look like.

- Withstand collision: Collision in hashing means the algorithm produces the same hash for two different values. SHA-256 has an extremely low probability of collision, and this is why it's heavily used.

All of these properties of the SHA-256 are the reason why blockchains are the way they are.

Let's understand the effect that these properties have by going over the following few points:

- Irreversibility translates into immutability in blockchains (transaction data, once recorded, can't be changed)

- Determinism translates into a unique, identifiable hash that can identify a user, wallet, transaction, token, or account on the blockchain (all of these have a hash)

- The avalanche effect translates into security, making the system extremely difficult to hack since the information that's encrypted can't be predicted by brute force (running multiple computers to estimate incrementally, starting with a hypothesis)

- Collision tolerance leads to each ID being unique and there being an extremely high mathematical limit to the unique hashes that can be produced, and since we require hashes to represent various types of information on the blockchain, this is an important functionality

In this section, we have seen how the properties of blockchains actually come from the hashing algorithms, and we can safely say that it's the heart and soul of a blockchain.

Transactions

Because of the previously mentioned properties of blockchains, storing financial data is one of the biggest use cases that blockchains are used for, as they have advanced security requirements.

A transaction is showcased through unspent cryptocurrency, or **unspent transaction output (UTXO)**. This refers to unused coins owned by individuals logged on the blockchain for transparency. It's essential to recognize that while UTXO is a key element in certain blockchains such as Bitcoin, it's not a universal feature across all blockchain platforms.

The following diagram helps us visualize all the fields in a transaction:

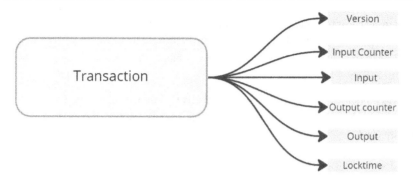

Figure 1.5 – The contents of a blockchain transaction

Let's go through all the fields that form a Bitcoin **transaction**:

- Version: This specifies which rules the transaction follows
- Input counter: This is the number of inputs in the transaction (this is just a count)
- Inputs: This is the actual input data
- Output counter: This is similar to the input counter, but it's for keeping a count of the transactions' output
- Output: This is the actual output data from the transaction
- Blocktime: This is simply a Unix timestamp that records when the transaction happened.

Initially, blockchains were primarily designed to record financial transactions within the realm of cryptocurrencies. However, as they evolved, blockchains demonstrated their versatility by finding applications beyond this initial purpose. Soon, we'll delve into these additional uses.

But for now, it is important to understand that when we mention transactions, it does not strictly mean financial or currency-related transactions. Rather, in modern blockchains, a transaction is *anything that changes the state of the blockchain*, so any program that runs or any information that's stored is simply a transaction.

Security

So, the main selling point for blockchains is that they're extremely secure. Now, let's understand why this is so:

- All the records are secured with cryptography thanks to the SHA-256 algorithm.
- The records and other blockchain data are copied to multiple nodes; we will learn about this in the *Peers, nodes, validators, and collators* section. Even if the data gets deleted in one node, it doesn't mean that it's deleted from the blockchain.

- To participate as a node in the blockchain network, requiring ownership of private keys is essential. Private keys and secret codes known only to you, grant access to control your cryptocurrency holdings, sign transactions, and ensure security. Possessing private keys safeguards your digital assets and enables engagement in network activities.

- Nodes need to come to a consensus on new data to be added to the blockchain. This means bogus data and corrupted data cannot be added to the blockchain, as it could compromise the entire chain.

- Data cannot be edited on the blockchain. This means the information you have stored cannot be tampered with.

- They're decentralized and don't have a single point of failure. The bigger the network or the more decentralized the network, the lower the probability of failure.

We will learn about nodes, decentralization, validation, and consensus later on in this book, and all these points will be clearer.

Storage versus compute

Bitcoin introduced blockchain for the storage of financial transactions, but Ethereum took things a bit further and helped us imagine what it could be like if you could run programs on a blockchain. Hence, the concept of smart contracts was created (we will dig deeper into smart contracts later in this chapter, but you can think of them as code that can run decentralized on the blockchain).

Independent nodes could join a network for the blockchain and pool their processing power in the network.

According to Ethereum, they're building the biggest **supercomputer** in the world. There are two ways to build the biggest supercomputer— build it centralized, where all machines will exist centrally in one location, or build a decentralized version where thousands of machines can be connected over the internet and divide tasks among themselves.

Ethereum enables you to process programs on the blockchain. This means anyone on the internet can build a smart contract and publish it on the blockchain where anyone else across the world can interact with the program.

This is the reason we see so many startups building their products on the Ethereum chain. After Ethereum, blockchains such as Solana, NEAR, and Polkadot have taken this idea much further and brought many new concepts by improving on Ethereum. This book is going to deal with all three of these blockchains.

Exploring the backbone of blockchains

This section is a deep dive into what makes blockchains so special. We will cover topics such as *decentralization*, *forking*, and *mining*, and we will understand how peers interact in a network and how the blocks are validated. Let's dive in.

Decentralization

From a purely technical standpoint, Web 1.0 started with a client-server architecture, usually monoliths. When traffic and data started increasing, the **monolithic** architecture couldn't scale well. Then, with Web 2.0, we had concepts such as **microservices** and **distributed systems**,which helped not only scale systems efficiently but also enhanced **resilience** and **robustness**, reduced **failure** instances, and increased **recoverability**.

The data was still centralized and private and the systems were mostly centralized, meaning they still belonged to a person/company and admins could change anything. The drawbacks were the following:

- A failure at the company's end took the system down

- Admins could edit the data and block users and content from platforms

- Security was still not prioritized, leading to easy data hacks, although this could vary depending on the company's approach to safeguarding information

- All the data generated on the platform belonged to the platform

- Content created and posted on a platform became the property of the platform

Web 3.0 ushers in a new age of decentralization that is made possible with blockchains where the entire blockchain data is copied to all the nodes. But even distributed systems had nodes and node recovery, so the question is, how is this any different?

Well, in the case of distributed systems, the nodes still belonged to the centralized authority or the company that owned the platform, and nodes were essentially their own servers in a private cloud. With decentralized systems, the node can be owned by another entity, person or a company other than the company that developed the blockchain.

In fact, in a blockchain network, having nodes owned by different companies is encouraged and this increases the *decentralization* of the network, meaning there is no real owner or authority that can block content, data, or users out and the data is accessible to all the nodes since all of them can store a copy of the data.

Even if one node goes down, there are others to uphold the blockchain, and this makes the system highly available. Advanced communication protocols among the nodes make sure the data is consistent across all the nodes.

Nodes are usually monetized to stay in the network and to uphold the security of the network (we will read more about this in the next section). Nodes also need to come to a consensus regarding the next block that's to be added to the chain. We will also read more about consensus shortly.

Peers, nodes, validators, and collators

In this section, we will further build upon the knowledge we have gained in the past few sections. A blockchain does not exist in isolation; it is a **peer-to-peer network**, and all full nodes save the complete copy of the blockchain, while some blockchains also permit other types of nodes that maintain state without necessarily possessing a full copy.

In the following diagram, we see this in a visual format:

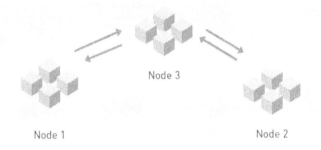

Node 3

Node 1   Node 2

Figure 1.6 – Multi-node networks

So, let's dig a layer deeper. Nodes are listening to **events** taking place in the network. These events are usually related to transactions. It is important to reiterate that a transaction is anything that changes the state of the system.

As we know, a block contains the information of multiple transactions.

The following diagram shows a block with some example transactions:

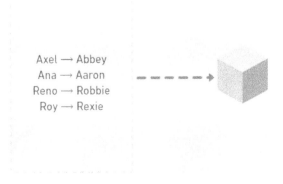

Figure 1.7 – Transactions finalized to a block

Once a new block is added by a node, which is known as mining, this new event is advertised to the entire network. This is visually represented in the following diagram:

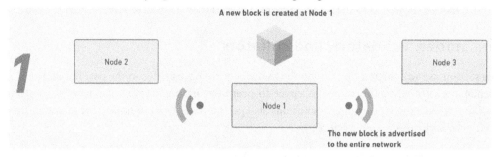

Figure 1.8 – The created block is advertised

Once the new block is advertised, the rest of the nodes act as *validators* that confirm the outputs of the transactions once the block has been validated by the rest of the nodes. The nodes come to a *consensus* that yes, this is the right block that needs to be added to the chain. We can visualize this with the help of the following diagram:

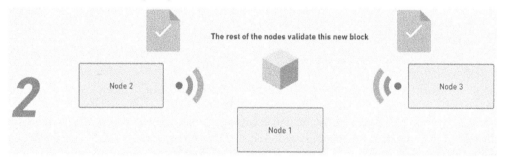

Figure 1.9 – Other nodes validate the block data

The new block is then copied to the rest of the nodes so that all of them are on the same page and added to the independent chains being maintained at each node. This can be seen in the following diagram:

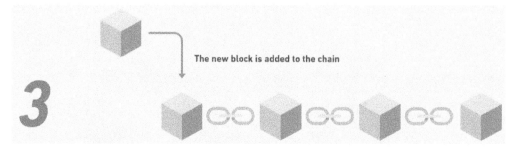

Figure 1.10 – A block gets finalized

Once the blocks are added to the node, and the blockchain at each node is updated on any other node. Another block could be listening to all the new transactions that have happened, and these are then collated onto a block and the entire process then repeats.

The following criteria vary from blockchain to blockchain in terms of the following:

- The number of transactions that the block will store
- The mechanism that nodes use to collate the transactions (time-based or number-based)
- The validation mechanism
- The consensus mechanisms

Contemporary chains improved upon the Bitcoin and Ethereum blockchains by varying and innovating on either all or some of these criteria, but the consensus mechanism is something that is most often innovated upon. This is done to try and save the time required for new nodes to be added and copied by the entire network, which is what really slows down the network.

We learned earlier that the nodes need to be incentivized to stay in the network and keep adding the blocks to the chain. In chains such as Ethereum, this is achieved using **gas fees**, which are simply small fees that users pay to carry forward their transactions. We know that blocks can contain only a few transactions, and if the users want their transactions to get priority, they need to pay gas fees.

The gas fee depends on what other users are willing to pay to get their transactions forwarded; the higher the gas fee, the higher the chance of getting your transaction accepted. Think of gas fees as the rent that the nodes get paid for the users to use the nodes' processors to process and validate their transactions. The words *peers* and *nodes* are used interchangeably, and *validators* and *collators* can also be used interchangeably depending on the blockchain you are on.

Consensus

In the last section, we learned that a node listens to transaction events, collates these transactions, and creates a block. This is called **mining**. After a block is mined, other nodes need to validate it and come to a consensus.

In this section, we want to peel the layers of consensus to understand it deeply. Understanding the mechanics behind some popular consensus mechanisms will help us to learn by running through actual examples, rather than learning in an abstract way. So, let's understand some of these concepts:

- **Proof of work (PoW)**: Nodes need to solve a particular cryptography problem (we will look at this in detail in the *Mining* section), and the node with the highest processing power is usually able to solve faster than others. This keeps the system decentralized but increases electricity consumption by a huge amount. It's not considered to be very efficient and is even considered bad for the environment, as it increases power wastage since all the nodes are up against each other trying to solve the problem. Examples are Bitcoin, Litecoin, and Dogecoin.

- **Proof of authority (PoA):** This is a consensus mechanism in blockchain where transactions and blocks are verified by identified validators, typically chosen due to their reputation or authority. Unlike energy-intensive mechanisms such as PoW, PoA offers efficiency by requiring validators to be accountable for their actions. It's commonly used in private or consortium blockchains, ensuring fast transactions and reducing the risk of malicious activities. However, PoA's centralized nature may raise concerns regarding decentralization and censorship resistance compared to other consensus methods.

- **Proof of stake (PoS):** Nodes need to buy *stakes* in the network—basically, they buy the cryptocurrency native to the network. Only a few nodes with a majority stake get to participate in the mining activity in some cases. This is highly power efficient, and this is the reason why Ethereum recently switched from PoW to PoS. However, it is considered to be less decentralized, as only the nodes with enough resources get to add the next blocks and it can be seen that some big players have been slowly taking ownership of the majority of the network since Ethereum switched to PoS. The main benefit of PoS is that since nodes have a stake in the system, they are de-incentivized to add unscrupulous blocks to the chain. Since the copy of the chain exists with all the nodes of the entire network, the nodes are running the *software* of the blockchain where the output hashes need to be consistent with the rest of the chain. Hence, when a node tries to add the wrong block, the rest of the nodes do not validate this block, and if such a scenario takes place, these nodes are then penalized where the amount of native cryptocurrency owned by the node that is taken away can differ depending on the seriousness of the violation. Generally, this penalty entails a partial loss of funds rather than a complete forfeiture of all holdings. Some examples are Cardano, Ethereum, and Polkadot.

- **Proof of burn (PoB):** Burning is a process where cryptocurrency is sent to a wallet address from which it's irrecoverable. The nodes that can burn the highest amount of cryptocurrency get to add a node. Miners must invest in the blockchain to demonstrate their commitment to the network. Even though PoB is the most criticized consensus model, it can actually be highly effective for some blockchains that want to ensure deflationary tokenomics. Slimcoin is an example of PoB.

- **Proof of capacity:** In this consensus mechanism, the nodes with the highest storage space get to add a node. This means that the nodes that partake in the network can use their hard drive space to compete with each other to win the mining rights. An example is Permacoin.

- **Delegated PoS:** Participants in the network, such as end users buying cryptocurrency, can stake their coins in a pool, and the pool belongs to a particular node that can add blocks to a chain. The more tokens you stake, the bigger your payout. Examples are EOS, BitShares, and TRON.

In this section, we've developed a rich understanding of consensus mechanisms, and this will help us throughout the book, especially while building the blockchain.

Mining

By now, we have a very basic idea of what **mining** is and why it's necessary. In this section, we will dive into the specifics of mining. Mining happens quite differently in different consensus mechanisms. We will look at mining for the two major consensus mechanisms: PoW and PoS. For instance, in PoS, let's consider the example of Ethereum 2.0, where validators are chosen to create new blocks and secure the network based on the amount of cryptocurrency they hold and are willing to "stake" as collateral.

In a PoW blockchain, to add a block to the blockchain, a cryptographic problem needs to be solved. The node that comes up with the solution first gets to win the competition. This means that nodes with the highest computational power usually win and get to add a block.

The blockchain's cryptography challenge adjusts in complexity over time to ensure consistent `block` creation. Nodes predict a specific hash, focusing on a segment that aligns with the existing blockchain, maintaining chain coherence.

Nodes employ a nonce, a unique value, to address the challenge. Incrementing from zero, this value is adjusted until a matching hash is computed, pivotal for generating a valid hash in line with the network's rules.

Solving the cryptographic problem validates transactions and creates new blocks. A successful node broadcasts its solution, swiftly verified by others. The first to find a valid solution is rewarded with newly minted cryptocurrency, incentivizing participation and bolstering network security.

The following diagram shows the different fields that add up to produce a hash:

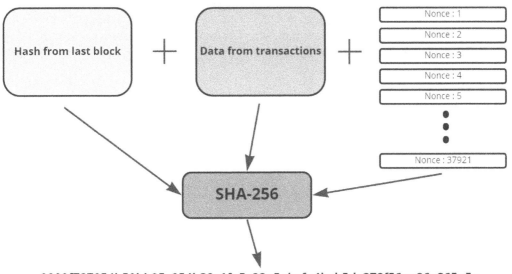

Figure 1.11 – All the data that makes up a hash

Now, this means the following:

- This can only be solved with brute forcing, iterating from zero to a particular number, and cannot be solved *smartly*

- All the nodes in the network compete with each other regardless of whether they ever win, and this means a lot of computational energy gets wasted

- Nodes need to keep upgrading their computational power to win the competition

Now that we understand what mining is and how it works, it's time to learn about forking—an important blockchain concept.

Forking

There is one small detail about blockchains that we have talked about but haven't discussed in detail yet, and that's **immutability**. In one of the earlier sections, we learned how SHA-256's properties translate into immutability for blockchains. This means all transactions that happen on-chain are immutable, and tokens once sent from one account to another cannot be reversed unless an actual transaction is initiated from the second account.

In traditional payment systems, this is not the case. If money is sent to the wrong account by mistake, this can be reversed, but this feature has been manipulated by centralized authorities and therefore immutable transactions are valued highly.

Let's take as an example the **decentralized autonomous organization (DAO)** attack in 2016 that led to $50 million being stolen from the Ethereum blockchain due to a code vulnerability. The only way to *reverse* this was to create an entire copy of the chain where this particular transaction didn't take place. This process of creating a different version chain is simply called *forking*. This event divided the blockchain between Ethereum and Ethereum Classic.

The following diagram demonstrates what forking looks like:

Figure 1.12 – Forks in a blockchain

Forking also comes into use when rules for the blockchain need to be modified. Traditional software gets *upgraded* and new updates and patches are applied, whereas the way to upgrade a blockchain is to *fork* (though some blockchains such as Polkadot have invented mechanisms to have forkless upgrades).

Forks typically occur intentionally, but they can also happen unintentionally when multiple miners discover a block simultaneously. The resolution of a fork takes place as additional blocks are appended, causing one chain to become longer than the others. In this process, the network disregards blocks that are not part of the longest chain, labeling them as orphaned blocks.

Forks can be divided into two categories: soft forks and hard forks.

A **soft fork** is simply a software upgrade for the blockchain where changes are made to the existing chain, whereas with a **hard fork**, a new chain is created and both old and new blockchains exist side by side. To summarize, both forks create a split, but a hard fork creates two blockchains.

Permissioned versus permissionless

Blockchains can be permissionless or permissioned depending on the use case. A permissionless blockchain is open to the public with all transactions visible, but they may be encrypted to hide some crucial details and information if required. Anyone can join the network, become a node, or be a validator if the basic criteria are met. Nodes can become a part of the governing committee as well once they can meet additional requirements, and there are no restrictions on who can join the network. You can freely join and participate in consensus without obtaining permission, approval, or authorization.

Most of the commonly known blockchains, such as Ethereum, Solana, and Polkadot, are all permissionless chains and are easily accessible. Their transaction data is publicly available. So, a perfect use case for permissionless chains is hosting user-facing and user interaction-based applications.

Permissioned chains have gatekeepers that define a permission, approval, or authorization mechanism that only a few pre-authorized nodes can operate. So, to be a part of the permissioned blockchain network, you may need a special set of private keys and may also need to match some security requirements. Since the nodes copy the entire data of the chain and are also involved in adding blocks to the chain and being a part of the governing committee for the blockchains, some use cases where data and information need to be kept private can use permissioned chains.

The following diagram shows the difference between a public and a private blockchain network:

Public versus Private Blockchain Network

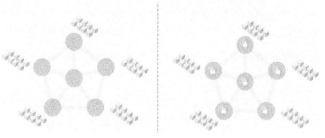

Figure 1.13 – Permissioned versus permissionless chains

Governments, institutions, NGOs, and traditional corporations have found plenty of use cases for permissioned chains, where only a few actors trusted by the centralized authorities are permitted to join the network. Permissioned blockchains also have multiple business-to-business use cases and may be centrally stored on a single cloud provider.

Blockchains help us decentralize computing and resources, and we have been using the word *decentralization* quite often. In the next section, we will understand the concept of decentralization in more depth.

Understanding decentralization

Decentralization is the guiding principle for Web 3.0. It's designed to create a win–win environment for the builders of a platform, the people that build *on* the platform **decentralized applications (dApps)**, and the people that interact with the platform (users of dApps).

Let's try and understand why decentralization is so important. In 2013, Twitter had a centralized developers platform where developers could use their APIs to build apps on the Twitter platform. A few years later, Twitter stopped the API support and also brought in a few restrictions, and every few months, the API's terms and conditions would change. This affected many app developers who were either banned from the platform due to the restrictions or were unable to stay up to date with the changing terms for API usage.

Similarly, Facebook had an app developer program as well, which many developers built their apps with. However, developers faced similar problems here as well, and this problem is quite common wherever a centralized platform is involved. Play Store and App Store can ban any app from their platform, and Amazon can decide which sellers can sell and Uber can decide which drivers get more rides.

The issue is not just about getting banned from the platform and the policy changes, but it's also about monetization. For example, the Apple App Store can take about 30% of the entire revenue from app developers. To prevent institutions, banks, and governments from curbing the freedom of individuals and communities, decentralization is a popular solution that ensures everyone gets a voice and a few owners of the platform do not end up controlling the entire platform.

It's shared ownership where the ownership of the platform is not held closely by the founding team or the committee; rather, it belongs to the community at large where each user can hold tokens and gets a say in the system. We will read about this further in the *DAOs* section.

A blockchain network implements decentralization in a highly efficient manner, and this is why it's the primary technology for a decentralized use case.

So, now that we have a clearer understanding of decentralization, let's dig into some of the concepts that are closely related with decentralization that make it possible.

Replication

Replication is at the core of decentralization. Multiple copies of the same data exist at different participatory nodes. If one of the copies gets corrupted, there are other copies to reference the final state of the data. If one of the participatory nodes fails completely, there are mechanisms by which a new node can be appointed to take its place, and this ensures that the entire network can't go down easily.

Even if the network does go down, recoverability is way easier when the data is replicated as opposed to having a centralized store of data. In decentralized systems, both the data and the power to govern the system can also be decentralized. This means that multiple nodes that have access to and copies of the data or that simply have a *stake* in the network can come to a consensus on the new data that gets added and how the network behaves/can be improved.

This leads us to the next important topic of governance.

Governance

Every blockchain follows a unique way of **governance**, but, in general, blockchains are governed by token holders that can be users, developers, investors, the founding team, or nodes. Governance tokens are issued to reward the loyalty of token holders and are usually issued when they have a high stake in the network. Owners of governance tokens can propose changes and vote for the implementation of those changes to the protocol.

Each governance token is equal to one vote, so this leads to a very fair and just system, as the holders with more stake in the network get more influence. The details of each vote are publicly available for all to see. Blockchain networks sometimes suffer from the *whale* problem, where some network participants end up holding most of the tokens. This can lead to the network becoming more centralized than expected. In most cases, whales end up being the founding team, and votes are cast to push the agenda of the founding team. This ends up centralizing the network.

For a blockchain network to be more effective, it is important to ensure a fair distribution of tokens to ensure decentralization. Governance usually aims to ensure that the blockchains update continuously and the protocols align with the right direction. However, it's worth noting that in certain cases, due to the influence of governance token holders, updates might not occur, potentially causing a divergence from the intended path.

Cryptocurrencies and gas fees

In the earlier sections, we read about cryptocurrencies and gas fees, and now that we have preliminary knowledge of blockchains, it's a great time to understand this topic even better. We will look at cryptocurrencies and gas fees from a slightly different perspective to get a more wholesome understanding of the topic.

Cryptocurrency is simply a currency that's cryptographically secure and is backed by a blockchain. Since a **hashing algorithm** is used to create hashes in a blockchain and this secures the blockchain cryptographically, using blockchains to store currency-related financial information makes a currency a *crypto*currency.

Blockchain can be thought of as a distributed ledger, and therefore it is referred to as **distributed ledger technology** (**DLT**). This is where a ledger is being maintained for the transactions that are being recorded, but also, multiple copies of the ledger exist with different nodes.

Since multiple copies of this ledger exist with multiple different nodes, the nodes must commit their storage space for this ledger. We also learned that when the blocks are mined, substantial processing power is utilized (PoW). This implies that nodes also need to commit processors along with storage space.

Since in this section we have focused on decentralization, it is important to understand that the higher the number of independent nodes that maintain a copy of the ledger and participate in the mining of new blocks, the more decentralized the network is. Hence, it makes it more trustworthy for developers and users to either build on this or interact with it respectively.

To keep the nodes incentivized to dedicate substantial storage and processing resources to the network, we need to pay them a fee or reward them with new tokens, which is simply gas fees. We've read a little about this in the previous sections.

The gas fee is paid out to the nodes in simply the *native* cryptocurrency of the blockchain. For blockchains to be effective, a coin that represents some limited digital asset is introduced. Please note that coins and tokens have various similarities, but they also have differences, and we will learn about them in one of the later sections.

In this section, we have filled in many gaps in our knowledge. It is now clear to us that for blockchains to be effective, they need to be decentralized, and for them to be decentralized, they need to have more nodes in the network. If we'd like the nodes to dedicate resources, they need to be incentivized, and if they want to be incentivized, we need them to have some monetary benefit that can be created with the help of a coin. This is a cryptocurrency, as it is secured by a blockchain for which security originates from a hashing algorithm. This is a cryptography hashing algorithm, hence the term *cryptocurrency*. While the compensation given to nodes for their role in maintaining the system and keeping the records safe is referred to as a gas fee.

Decentralized platforms

In traditional Web 2.0 platforms, the data, as soon as uploaded or created by the users, legally belongs to the platform. Every time a user logs into a traditional Web 2.0 platform, the request for authentication is sent to the centralized server with a centralized database, both of which are entirely owned by the platform. The platform can decide if a user can stay on the platform or is asked to leave. This centralizes ownership and control.

Now, the platform has complete ownership rights over all the data that is generated by the users that use the platform.

Since the platform is the owner of the data, the platform is responsible and is also incentivized to keep the data secure.

But since the data is users' data, it's actually the users that stand to lose the most when the data actually gets hacked or stolen.

We have seen this pattern multiple times in the past; platforms get hacked often and user data is stolen and sold by hackers for a small profit.

So, even though the platform is incentivized to secure the data, the platform doesn't really stand to lose much when it's stolen, and therein lies the problem.

In addition to this, the platforms also sell this data themselves to third parties or use the data for advertising to these users, sometimes without consent. This further deteriorates the user experience and trust in the platform.

Web 3.0 re-imagines this model; the data instead stays with the user and they can choose to hold it in their own secure wallet and use this wallet as an authentication mechanism on different platforms. This means they can take their data with them.

This means that there's no central ownership over the data and no control over the rules of the platform, as the data exists with multiple nodes. Multiple nodes can vote for governing the network. This creates a healthy ecosystem for developers to come and build decentralized applications on the platform that will further attract even more users. At the beginning of this section, we read that decentralization creates a win–win solution for all the parties involved, but we hadn't understood why. Now we know the reason.

The whole purpose of Web 3.0 and decentralization is to create healthy, secure ecosystems where more people trust the system inherently, not because they trust the central authority that controls the system but because they trust the mechanism and processes around how the system is set up.

So, to summarize, in Web 2.0, the tagline was *Don't be evil* (this is Google's famous tagline), meaning that even though the platform owns the data and is responsible for securing it, everything depends on the trust that users have in the platform. This trust has been exploited multiple times, whereas in Web 3.0, the motto is *Can't be evil* since the system is set up to not have a central authority. This means it's built on a zero-trust system where users don't have to trust the platform since the system takes care of the fact that the platform never ends up owning users' information.

Tokens versus coins and ICOs

We learned that cryptocurrencies are created to incentivize the nodes in a network to dedicate their storage and processing resources. These native cryptocurrencies running on original chains and their forks are considered coins.

All native cryptocurrencies, without exceptions, run on their own blockchains and have an intrinsic value that's linked to multiple factors, with the efficiency and security of the underlying blockchain being two of the primary ones.

Tokens, on the other hand, don't need an underlying blockchain and can be used by the following:

- Technology projects to raise capital
- Technology companies to represent digital assets
- Corporations, institutions, and individuals to represent a real-world resource in the digital world

Now, let's discuss these three use cases in order.

The first use case is for technology projects when the technology required to be built is complex or advanced and will require some runway expenses for the team to research and build the project.

For early-stage and research-driven tech projects, raising capital from the general public is not an option, as that space is highly regulated.

So, instead of raising capital from traditional venture capitalists, such tech projects can leverage **initial coin offerings** (**ICOs**) to raise capital wherein a particular amount of tokens are generated. A small percentage of these are distributed among the founding team, developers, and some early-stage users. After this distribution, the majority that are left are purchased by the next set of users.

When users buy the token, they are essentially buying into the vision of the tech project and funding the runway required to build it. This process can be likened to a crowd-funding process.

The second use case is related to representing digital assets, such as the following:

- **Smart contracts** to indicate property ownership, mortgages, and digital identity
- **Non-fungible tokens** (**NFTs**) to represent art, music, in-game assets, and so on

We will learn about both in the next section.

The third use case is related to representing real-world, physical resources and assets in the digital world. Tokens have the perfect use case for this. Real estate assets, crops, medicines, and so on can be tokenized in the digital world and represented on the blockchain. They can then be traded between users, each trade representing an actual transfer of value between the users.

The presence of blockchain ensures that the immutable ledger is maintained and updated, and the immutability and cryptographic security provided by the blockchain bring trust in trading such physical assets digitally. This opens the possibility of cross-border transactions and trades since such transactions are scalable digitally.

Smart contracts and NFTs

In the *Tokens versus coins and ICOs* section, we learned how tokens can be used to create smart contracts and NFTs. Now, let's discuss what these are in a little more detail. It is important for us to understand them because as a blockchain developer, you will be creating smart contracts and even NFTs all throughout your career.

Let's first tackle smart contracts. For a real-world, regular contract to take place, there is a governing, intermediary body such as a bank, legal authorities, or the government. This is basically an entity that has some enforceable power in the real world in case the parties signing the contract do not hold up their end of the bargain.

In the case of smart contracts, there is no intermediate party. Instead, there's just a program/computer code that runs on the blockchain, and this code is cryptographically secure and immutable (the two properties that are induced due to the SHA-256 algorithm).

The smart contract ensures that none of the parties can back out of the contract or make any changes, and it also ensures that the parties hold up their end of the bargain without involving any manual intervention or human beings and institutions as authorities and witnesses.

Irreversible, immutable smart contracts are objective in nature, whereas regular contracts can be highly subjective in nature because the differences in interpretations of the contract usually lead to discord and conflict.

We mentioned that a smart contract runs on a blockchain, but how does that actually work? A smart contract gets its own hash, which is used to recognize it, and once the smart contract is executed in a particular way with different values, you get a different **application binary interface** (**ABI**) (a representation of your contract to be called from the external world). This is what's stored onto the blockchain.

Let's walk through a use case example to understand how this actually functions. Let's say there are two parties involved. The first is a corporation that wants to transfer some stocks to an individual who works in that firm but the firm wants these stocks to be transferred based on the number of months that the individual works at the firm rather than all at once. This process can be completely automated with a smart contract that can be integrated to send push events to the stock vesting service based on a CRON script that runs every few months based on the frequency and schedule mentioned in the contract.

Now, this brings in a lot of trust between the corporation and the employee who will be given the stocks since there is no intermediary who can be swayed. There is just a computer program that gets executed and is immutable. Now that we have a solid understanding of smart contracts, let's dive into NFTs, which are simply an extension of smart contracts.

In the previous section, we read that NFTs can be used to represent art, music, and in-game assets. Now, let's understand how that works. **NFTs** are **non-fungible tokens**, meaning they are different from regular tokens that are fungible.

Although each NFT is distinguishable from another, they *can* be exchanged with each other. The fact that they are distinguishable makes them non-fungible, not their exchangeability.

We can convert images such as a JPEG file into an NFT by assigning it a hash, and this process is called *tokenization*. Each NFT token has a unique hash and it cannot be exchanged since each NFT is associated with only one unique hash.

But we know that a crypto token is fungible and can be exchanged (if you have one crypto token and your friend has one crypto token and you exchange them, it doesn't matter because those tokens aren't unique, thereby making them fungible).

Since you can assign unique hashes to NFTs, this means you can represent real-world contracts in the digital world using NFTs. This is quite different from a smart contract, which isn't a real-world contract and has no authorities involved.

DAOs

In the previous sections, we learned how real-world assets can be represented in the real world with tokens and NFTs, but what if we applied the same concept to organizations and corporations?

Using blockchain, we can imagine a different future for organizations where the power will not be centralized. Currently, many issues can be pointed out in organizations that are centrally run by just a handful of people. Such organizations can sometimes work against the society and the community.

But here's a revolutionary thought: what if an entire community takes up the responsibility of running an organization? This means tokens would be issued to the people from this community and they would represent ownership. Some of these tokens would be governance tokens and would have more weight in the voting process.

An elaborate voting process would take place for critical decisions, and the voting weightage would be based on the tokens you hold.

All of the decisions resulting from the voting process would be made public and the rest of the community that has regular tokens and not governance tokens would be apprised of the updates.

The public would also decide to buy these tokens and take up a seat at the table for governance.

Building such a community-led organization would ensure that the decisions are always taken for the benefit of a very small percentage of society. Since all the decisions are available on a public blockchain, they are mostly irreversible, so all the promises made by corporations would need to be fulfilled.

This model of organization is known as a **decentralized autonomous organization** (DAO).

This is a great model to run a highly democratic organization, and it is thought that even governments can run like this.

This presents us with a completely digital way to run a real-world organization fully remotely. Currently, many companies are experimenting with this model. It is important to note that the DAO model is not currently completely free of issues and vulnerabilities, and the factors that affect organizational decision-making in the real world can also affect DAOs. It is an evolving model and something that's widely agreed upon as the future.

Non-censorable apps

Decentralization makes it extremely difficult to censor information since multiple copies of the same information exist, and it also makes it significantly more difficult to spread fake news and propaganda, as facts can be recorded on a blockchain. The facts are immutable, as the information cannot be changed later.

A significant number of dApp developers are focusing on creating Web 3.0 versions of apps such as Twitter, Reddit, and Quora where posts cannot be taken down by admins, firstly because there would not be any admins since the app would be community-driven and secondly because the information would be immutable so there would be no edit or delete functions. This enables true freedom of expression on an unprecedented scale.

So, now we have clarity on how information on apps can be resistant to censorship, we can go beyond by making websites and web apps censorship-resistant at the fundamental level.

This means that it shouldn't be easy to take down websites by just restricting traffic to them. This is made possible by decentralized storage platforms such as Arweave and EmbassyOS, where independent individuals can buy special physical storage devices.

These storage devices can be used to store the websites on their own devices rather than on a cloud server, making it significantly more difficult to take websites down.

Digital assets with real-world limits

Blockchains would only matter to the common man if they're able to affect the real world or our daily lives. So, let's learn about how blockchains model real-world limits and how blockchains can be used to represent real-world resources.

Up until now, we've never had a way to represent real-world resources in a digital format, but blockchain and tokenization enable this for us. It's important to discuss this, as it opens tons of possibilities, so let's walk through a few use cases to understand it better.

Let's say there's an apartment in Japan with 1,000 square meters of space. If you want to buy this apartment, you must buy it as a whole, and you also need to meet a lot of criteria, such as being a resident of Japan. But what if there was a way to not buy the entire apartment and buy only a fraction of the apartment? With tokenization, this is made possible.

This is how the process works: a tech enablement company acquires this apartment and tokenizes the square meters. Each token represents one square meter of the apartment, and this currency is restricted to 1,000 tokens (since the total space of the apartment is 1,000 square meters). This total limit cannot change since it's backed by a blockchain, and now these tokens can be purchased by anyone in the world.

This means that if you want to buy just two square meters of that apartment in Japan, you must buy two tokens. The benefit of this is that you can invest in the real estate market of Japan and the growth that Japan will see in the coming years by just buying the token. This opens the market up for multi-ownership models and removes long processes from the equation. Since these tokens can be traded, all trades will be recorded on the blockchain, and the latest ownership state will be maintained and updated. This same concept can be applied to many other things in the real world, and right now, we're just at the cusp of a new wave of tech enablement that may end up changing our reality for the better.

We now know that blockchains have a lot of use cases, and for blockchains to be truly adopted, we need blockchains that can process a lot of transactions and handle a lot of users. In the next section, let's look at how blockchains can be scaled to handle millions of transactions and users with ease.

Scaling the blockchain

Multiple nodes are involved in a blockchain network, and the number of nodes determines the decentralization coefficient, or the **Nakamoto coefficient**.

The Nakamoto coefficient measures the number of nodes required to maintain healthy decentralization in the network. This means the higher the number of nodes, the higher the Nakamoto coefficient.

Now, we know that all of the nodes in a blockchain network not only need to be aware of the transactions that are taking place in the network but also need to process these transactions and communicate with the other nodes to reach a consensus on the block to be finalized.

Since so many operations are taking place between the nodes in a network at any given point in time, it makes it difficult to scale transactions beyond a particular point. For example, the limit for Ethereum is 20 transactions per second. Many new blockchains are trying to solve this problem, and some companies are trying to solve this problem for existing blockchains such as Ethereum using some innovative solutions. In this section, our focus will be the problem of scalability.

In the next few sub-sections, we will cover the factors affecting blockchain scalability and some solutions to help scale blockchains with ease.

The blockchain trilemma

Before we dive deeper into scalability and look at the factors affecting it and its possible solutions, we first need to understand the blockchain trilemma, as just looking at scalability in isolation and solving for scalability does not provide us with an efficient solution.

Three different elements are highly desirable in blockchains: decentralization, security, and scalability. The blockchain trilemma states that as a blockchain network evolves, it becomes difficult to maintain all three of these traits, and usually there's an imbalance. One of the traits ends up being more dominant than the other two, and trying to enhance the others may end up weakening the dominant one.

The more nodes in the network, the higher the decentralization, but the scalability (number of transactions) goes down, as all nodes need to process these three elements and come to a consensus. Also, security can be more easily compromised since there are more points of entry.

So, to address scalability without affecting the other two elements and solving the problem effectively, we have a few options. We will go over them now.

Sharding

Blockchain sharding is a technique used to scale blockchain networks and improve their performance by dividing the network into smaller, more manageable components called shards. Each shard is like a separate blockchain with its own set of validators and transaction history, but they are all interconnected.

The primary goal of sharding is to increase the transaction processing capacity of a blockchain network by enabling parallel processing. In a traditional blockchain, every node in the network has to process and validate every transaction, which can result in bottlenecks as the network grows larger. Sharding overcomes this limitation by dividing the network into smaller shards, each capable of processing a subset of the total transactions.

Here's a simplified explanation of how blockchain sharding works:

- **Shard creation**: The blockchain network is divided into multiple shards, with each shard assigned a subset of accounts or addresses. For instance, one shard could manage transactions related to addresses starting with the letter A, while another shard takes care of addresses starting with the letter B, and so forth. However, when considering sharding methods, the approach described resembles range partitioning, where data is distributed based on predefined ranges or categories. However, there are alternative methods such as hashing. Hash-based sharding involves distributing data across shards based on the outcome of a hash function applied to the data, offering a different way to achieve load distribution and network efficiency.

- **Shard processing**: Each shard operates independently and processes transactions related to the accounts or addresses assigned to it. This allows for parallel processing of transactions within each shard, significantly increasing the overall transaction throughput.

- **Cross-shard communication**: Since transactions can involve accounts or addresses from different shards, a mechanism is needed for communication between shards. This is typically achieved through a cross-shard communication protocol, where transactions that affect multiple shards are coordinated and validated.

- **Consensus and security**: Each shard has its own set of validators responsible for validating transactions within that shard. This means that the consensus mechanism of the blockchain network must be designed to handle cross-shard transactions and ensure the overall security and integrity of the network.

- **Shard coordination**: To maintain the consistency of the blockchain across shards, some form of coordination is required. Techniques such as cross-links, where the state of one shard is included in the block of another shard, or periodic checkpoints can be used to synchronize the shards and maintain a consistent global state.

By implementing sharding, blockchain networks can achieve higher transaction throughput, lower latency, and improved scalability. However, sharding introduces additional complexity in terms of shard coordination, cross-shard communication, and consensus mechanisms. Designing an efficient and secure sharding solution is an active area of research and development in the blockchain space.

Interoperability

Interoperability is an interesting concept and has multiple implications. An indirect implication of interoperability can be scalability. Each blockchain has its own set of data, digital assets, and tokens stored on the chain, but if a user wants to move this data to another chain, there can be various migration and compatibility issues.

Interoperability is when you have different blockchain protocols built with different technologies and their own way of operations to work together and exchange resources and assets seamlessly.

A lot of companies are trying to solve the interoperability problem by building *bridges* that make it possible to transfer assets across multiple blockchain networks. This also means that if a blockchain is unable to scale beyond a particular point in terms of storage of assets and transactions, these can be shipped off to other chains.

Consensus for scale

The developers of Solana figured out that the consensus mechanism of the chain itself can be a factor that limits the scalability of the network.

Solana has implemented a novel **Proof of History** (**PoH**), which is used in tandem with **practical Byzantine Fault Tolerance** (**PBFT**). PBFT addresses Byzantine Faults, where malicious nodes can disrupt consensus. By integrating *PoH*, which cryptographically validates time passage and event order, Solana streamlines the chronology crucial for Byzantine Fault Tolerance. However, this integration adds complexity, as each node needs to execute sophisticated software for consensus participation.

Using *PoH* enables Solana to theoretically process 65,000 **transactions per second** (**TPS**) (currently, it is around 3,000 TPS) as opposed to the 20 TPS provided by Ethereum. This proves that innovating with different consensus mechanisms can enhance the speed and scalability of a network.

Parallel processing

Many blockchain systems lack parallel processing, hindering scalability and speed. Without this capability, transactions are processed sequentially, causing bottlenecks and slower network performance. This limitation restricts their ability to handle a high volume of transactions efficiently, ultimately impeding widespread adoption and real-world applications of blockchain technology.

Newer blockchains such as Aptos and Sui, developed using Rust, have introduced a groundbreaking Layer-1 scaling solution known as **parallel processing**. This innovation significantly enhances scalability by allowing multiple transactions or tasks to be executed simultaneously within individual blocks. Unlike Layer-2 solutions that build upon existing blockchains, Layer-1 solutions such as parallel processing directly optimize the blockchain's core protocol. This approach fundamentally increases transaction throughput and network efficiency, paving the way for nearly limitless scalability and improved performance, thus addressing a major limitation of traditional blockchain systems.

In older blockchains, all the present nodes in the network process all transactions individually. The new transactions that are recorded on the blockchain are all present in the **mempool**. All of the nodes create new blocks using the same transactions. So, this means the more nodes in the network, the more decentralized the network is.

But this does slow down the network because instead of leveraging the processing power of the nodes in the network to divide and process the blocks to enhance speed, all the nodes are essentially processing the same transactions and a lot of computational energy is used up.

Aptos (the new blockchain we talked about) processes pending transactions from the mempool. At the same time, the nodes in the network divide these transactions among themselves and process different transactions instead of processing the same transactions. This makes the process way more efficient and highly scalable at the same time since the more nodes you add, the more transactions you can theoretically handle. Adding more nodes speeds up the network rather than slowing it down (as in the case of older blockchains).

Layer 2s and side chains

Sharding, consensus, and parallel processing are all **Layer-1 scaling solutions**. What if, instead of trying to scale at Layer-1, which may require us to make changes to the blockchain's architectural structure, we try to solve the blockchain trilemma by building on top of the blockchain on Layer-2? It's important to note that all the layers are imaginary and the terminology is used for better understandability.

An example of a **Layer-2 solution** is a sidechain, which is essentially a separate blockchain connected to the main chain. It's set up in a way that assets can flow between the chain flawlessly. The biggest difference is that sidechains can be configured with different modes of operation and rules, which can make them way faster than the main chain.

Transactions can be shipped off to the side chain for faster speeds, but they may still be verified by the main chain once the side chain sends the output of the transactions back to the main chain.

ZK rollups and optimistic rollups

Rollups are another highly popular Layer-2 scaling solution for blockchains. They take the transactions off-chain and, at the same time, ensure storage on the main chain for high security.

The way they do this is by compressing the transaction data to a great extent so that it becomes a fraction of the size and can be stored very easily on-chain. This results in significant throughput enhancement. Prominent examples of this approach include Arbitrum (an optimistic rollup) and PLONK (a zero-knowledge rollup). These rollups employ different techniques but share the common goal of enabling seamless and secure scaling, making them pivotal in advancing the capabilities of blockchain technology.

Now that we have understood rollups, it's time to talk about the two types of rollups: **zero-knowledge (ZK) rollups** and **optimistic rollups**. Optimistic rollups don't need to provide any proof when sending the compressed transactions to the main chain, whereas ZK rollups need to submit cryptographic validation proof.

Since with optimistic rollups there is no validation proof being submitted, they are essentially operating on the assumption that the nodes are not going to submit any fraud transactions, and this is why these are called *optimistic*. On the other hand, ZK rollups assume that they have no knowledge of the type of transactions that can be submitted, and thus comprehensive crypto validation is required.

Now that we have learned about blockchain basics, let's now understand how smart contracts work.

Introducing smart contracts

Interacting with blockchains involves a dynamic interplay between external code and the blockchain's internal architecture, particularly smart contracts. Smart contracts are self-executing contracts with predefined rules that automatically execute actions when specific conditions are met. External code refers to applications, scripts, or software components running outside the blockchain network.

To bridge the gap between external code and blockchain functionality, most blockchain platforms offer **remote procedure call** (**RPC**) APIs. **Application programming interfaces** (**APIs**) facilitate communication and interaction between distinct software applications, enabling them to work together harmoniously.

An API defines a set of rules and protocols that govern how software components should interact, making it easier for developers to use functionalities provided by another system without needing to understand its internal workings.

Think of an API as a waiter taking orders in a restaurant. Customers (developers) interact with the waiter (API) to request specific dishes (functions or data) from the kitchen (the system providing the service). The waiter conveys the order to the kitchen, brings back the dishes, and serves them to the customers. The customers do not have to know how the kitchen operates; they just need to know how to communicate their orders effectively to the waiter.

Similarly, in the context of blockchain platforms such as Ethereum, RPC APIs act as intermediaries between external code (applications) and the core blockchain software. They provide a standardized way for external code to send requests for actions, data retrieval, or other operations to the blockchain. The blockchain's core software processes these requests and sends back the relevant information or results.

For instance, if developers want to retrieve the balance of an Ethereum address, they can use an **RPC API** call to request that information from the Ethereum network. The API handles the communication between the external code and the blockchain's internal systems, abstracting away the complexity of direct interaction.

The future of the adoption of blockchains

In this section, we will learn about the real-world implications of blockchain technology, how it will affect various industries and its effects on social and cultural aspects.

Industries disrupted

Blockchains provide a secure representation of real-world assets in the digital realm, thanks to their cryptographic security. This introduces a revolutionary perspective across all industries.

The following use industries are well covered by blockchains:

- Banking and finance
- Healthcare and medicare
- Supply chains and warehousing
- Governance and policy-making
- NGOs, associations, and institutions

Many of these industries are getting disrupted by blockchain technology simply because it adds trust, security, and decentralization, creating a win–win solution for all the parties involved in the ecosystem.

Sociocultural and economic changes

Apart from disrupting industries and bringing about technological change, efficiency, and effectiveness into the system, blockchains are also triggering social change, as it becomes easy to track whether all social strata are benefiting or if there are any individuals that are left behind. Applied with the right policymaking, they can ensure a reduction in economic disparity and wealth equality and ensure that economic welfare funds reach the intended audiences.

It is easy to see that applications and effects of blockchain technology go well beyond a few industries and can alter the way society and culture function at large, especially since blockchains allow the digital representation of real-world assets and thereby enable cross-border trading of these assets in a highly secure and efficient manner.

This is why it is important for engineers to gain expertise and a foothold in this new and upcoming technology.

Summary

In this chapter, we've gained deep insight into blockchains and how they operate. We covered many concepts that are important to understanding how blockchains truly work, such as immutability, forking, validation, transactions, nodes, the different consensus mechanisms, gas fees, and processing fees.

Then, we broke down how blockchains power decentralization and applications that are deployed to blockchains, such as DAOs, DeFi apps, NFT platforms, and so on. We also went through some topics such as scalability involving interoperability, consensus, and sharding. These are the hottest topics in blockchain technology since experts are working on them to try and scale blockchains for millions of users.

In the next chapter, we will understand what makes Rust the perfect fit for building blockchains and we will learn some Rust concepts hands-on before diving into building our own blockchain using Rust.

Rust – Necessary Concepts for Building Blockchains

Even though Rust is a new programming language, it's gaining popularity quickly since it makes the job of the programmer simple. With Rust, you get a simple promise – if your program passes the compiler's checks, it is most likely free of *undefined behavior*, in the sense that this reduces the chances of encountering unexpected bugs. However, it's important to note that no compiler can guarantee absolute freedom from all unexpected behaviors, especially in complex domains such as asynchronous and embedded code.

Rust is renowned for its speed and efficiency, often drawing direct comparisons with C and C++. It holds significant advantages over these languages, largely due to the proactive enforcement of rules by the Rust compiler. Unlike C and C++, where a multitude of rules exist and the onus is on the programmer to adhere to them, Rust assumes a more active role in rule enforcement. This fundamental difference is not just about the number of rules, but about the philosophy behind them. C and C++ operate under the belief that the programmer should have the knowledge and responsibility to follow best practices, while Rust embeds these best practices into its compiler's design, significantly enhancing the stability and safety of the code. Issues regarding memory management and pointer validity are taken care of (we will learn more about this in this chapter). Rust has all the tools to free programs of data racing.

Debugging Rust programs is much simpler because the potential consequences of a bug don't end up corrupting unrelated parts of your program. Also, Rust provides us with flexibility – in the sense that the applications for Rust are varied and wide. All of these advantages and many more make it one of the most loved languages of recent times. We will explore Rust in more detail in this chapter. You will gain insights into important Rust features, especially the ones we will need to build blockchains with. By the end of this chapter, you will be comfortable working with Rust and have developed a confident command over the basic concepts that find their way into many applications. The end goal is to equip you with enough Rust knowledge that you'll be able to understand the code in the following chapters.

In this chapter, we're going to cover the following main topics:

- Introducing Rust

- Rust's advantage for blockchains

- Learning basic Rust concepts

- Exploring intermediate Rust concepts

- Delving deep into advanced Rust concepts

Introducing Rust

Many blockchains have selected Rust as their go-to programming to write their core protocol on which the rest of the architecture is built. There are plenty of reasons for this and since our book is about Rust for blockchain development, we need to understand why Rust is so popular for the blockchain use case.

In the following subsections, our focus will be on understanding the reasons why Rust is a perfect fit for blockchains.

The benefit of being statically typed

Rust adheres to static typing principles, requiring an explicit declaration of variable types, which are resolved during the compilation process. This allows the compiler to check if a variable can do what it's supposed to, protecting against errors when the program runs.

In statically typed languages, the result or end product usually takes the form of a lower-level representation. Pre-compilation, the compiler possesses assurances for correctness and consistency regarding the structure of data entities, method availability, and more. Consequently, code crafted in such languages generally exhibits superior performance compared to interpreted dynamic languages. Dynamic environments necessitate heightened runtime checks for every instruction, incurring a performance overhead. With dynamic languages, many benefits are attributed, such as faster development cycles and less boilerplate code, but even these benefits are inherently present in Rust.

Having explored the advantages of Rust as a statically typed language, let's delve into its classification as a systems programming language and understand how its features align with this crucial role.

A dive into Rust's applicability as a systems programming language

Rust is highly suited for blockchains as it is a low-level language and a **systems programming language**.

A systems programming language allows us to write computer software that enables the programmer to interface with the hardware. Operating systems, firmware, compilers, and assemblers are examples of systems that can be built with a systems programming language.

Blockchains act as the infrastructure layer for decentralized applications and need to be highly efficient at using server resources. This is why systems programming languages with low-level control are a great fit.

At the same time, systems programming languages prioritize modularity, code reuse, and code evolution, which makes languages such as Rust highly effective.

The reliability of Rust

Rust is the most loved language for the seventh year in a row in the Stack Overflow programming languages survey (`https://survey.stackoverflow.co/2022/`). Most loved means that programmers consider it to be the most effective, reliable, and elegant language out of all the programming languages. However, this does not mean that it is the most widely used language and many languages rank better in terms of usage, number of developers, and ease of learning.

However, there are some reasons why developers prefer to use Rust if they're given a choice:

- **Detailed documentation**: The sheer number of resources available for Rust, such as its clear and detailed documentation, *The Rust Book* (`https://doc.rust-lang.org/book/`), the *Learn Rust By Example* (`https://doc.rust-lang.org/rust-by-example/`) website, and so on, make it easy to get started and gain mastery. Even though there is a learning curve to the language, having resources available makes the journey easier.

- **Community**: In specific cases when there is a lack of clarity on some aspects of the code and in case of errors that are not easy to figure out, developers can easily reach out to the Rust community on websites such as Stack Overflow and find several Rust projects created by the community on GitHub. There are plenty of pre-resolved issues already present that can be referred to.

- **Type safety**: TypeScript, developed to ensure type safety in JavaScript, addresses challenges similar to those that Rust tackles. While TypeScript adds type safety to JavaScript, preventing errors due to unsafe type casting, Rust inherently provides this in systems programming. This comparison is crucial as it highlights Rust's commitment to type safety, akin to TypeScript's role in JavaScript. Understanding TypeScript's impact on JavaScript's type safety helps us appreciate Rust's approach to ensuring safe, reliable code that's free from common type-related errors prevalent in other systems programming languages. Languages such as Rust are pre-built with type safety, and this makes it easy to find errors beforehand. **Type safety** ensures that the language doesn't allow an int to be inserted into a string at runtime and so on. A type-safe language maintains data truthfulness from the beginning to the end, thereby making them extremely reliable and stable for production-level code.

- **Data race-free**: To increase efficiency for the execution of complex programs, we can introduce parallelism and multithreading. Rust is a multi-threaded language that supports concurrency and parallelism natively. However, in multi-threaded programming, **data races** can occur when multiple tasks or threads access a shared resource without proper protection, leading to unpredictable results. For instance, consider two threads trying to update the same variable simultaneously, leading to inconsistent values due to a race condition. The occurrence of such an anomaly could entail a considerable duration to replicate the problem and an even lengthier span to ascertain the underlying source and rectify the flaw. Rust's memory management model steps in and extends to protect against data races when using shared-memory concurrency (we will learn about the memory management model in the *The Rust ownership memory management model* section).

- **Ahead-of-time (AOT) compiled**: This feature significantly reduces the amount of resources required and the amount of work needed to be done at run-time. AOT compilation is the process of compiling a relatively high-level language (such as Rust) to a low-level language (such as machine code). AOT is the opposite of **just-in-time (JIT)** compilation, which isn't considered as efficient as AOT.

- **Built on and encourages zero-cost abstractions**: A way to make a programming language faster is to take away the processing cost from runtime (when the program is running) and shift this cost to compile time (when the program compiles); Rust does exactly this. This implies that Rust programs run fast, but compile times can be slow and many developers who are trying out Rust for the first time complain about the long compilation times but aren't aware of the benefits this introduces to the speed of code execution.

Zero-cost abstractions take this concept a step further. This means that even if we add higher-level programming concepts such as generics, collections, and so on, these do not induce any runtime costs; only compile-time costs are included.

In the following subsections, we will discuss several features of Rust that make it one of the preferred languages for blockchain development.

The Rust ownership memory management model

Rust achieves the goal of eliminating memory errors without the runtime overhead of dynamic memory management (for example, garbage collection or reference counting). Rust does this with the help of the concept of *ownership*, which states that every piece of data stored in Rust will have an owner associated with it. Rust strictly tracks the lifetime of values, including references, to determine when a value can be deallocated and that no dangling references exist. This is how it's achieved:

- Every value has a single owner (for example, a variable, structure field, and so on) and the value is released (dropped) when the owner goes out of scope

- One mutable reference to a value may exist

- There may be any number of immutable references to a value and while they exist, they cannot be mutated

- All references must have a lifetime no longer than the value being referred to

Let us now focus on other capabilities of Rust.

Garbage collection

Garbage collection involves the automated process of reclaiming memory that the runtime is no longer using. In simpler terms, it serves as a means to eliminate unused objects. The purpose of garbage collection is to prevent a program from exceeding its allocated memory capacity or reaching a state where it can no longer operate properly. Additionally, it relieves developers from the manual burden of managing a program's memory, thereby reducing the potential for memory-related errors.

In contrast, Rust takes a distinct approach by forgoing the use of a garbage collector. Instead, it accomplishes these objectives through a sophisticated yet intricate type system (as discussed in the *The reliability of Rust* section). This methodology renders Rust exceptionally efficient; however, it also introduces a higher level of complexity to the learning and utilization of Rust.

This additional efficiency, which is due to there being no garbage collection, makes Rust a great fit for blockchains.

Speed and performance

Rust's performance is exceptional and rivals that of C and C++, languages renowned for their top-tier compilation-based performance. However, Rust distinguishes itself from these legacy languages by providing memory safety through its unique ownership-based memory management system (discussed in the *Rust ownership and memory management* section), as well as concurrency safety without significantly sacrificing execution speed. It excels in executing algorithms and resource-intensive tasks with remarkable efficiency, placing it on par with the performance of C++. Rust has a big advantage over C – Rust ensures the thread safety of all code and data, including those from third-party libraries, even if their authors didn't prioritize thread safety during development. Every element either adheres to specific thread-safety standards or is prohibited from being employed across threads. If I create any code that lacks thread safety, the compiler will identify the unsafe portions. Rust provides a guarantee against data races and memory vulnerabilities, such as use-after-free errors, even in multi-threaded scenarios. This assurance encompasses not only certain races that could be detected using heuristics or runtime analysis in instrumented builds but rather it encompasses all potential data races throughout the codebase. This is of utmost importance as data races represent the most critical form of concurrency-related bugs.

Futures, error handling, and memory safety

In this section, we'll look at some features of programming languages that enable asynchronous computing and processing to enable languages to be more efficient with resources and work in a non-blocking manner.

Just as JavaScript and Dart have the concept of async and await, Rust has *futures*. This feature makes it possible to await values of computational tasks that haven't finished yet but still enables developers to handle these values that will be present to them in the future.

Futures in Rust represent values that are eventually computed from time-consuming operations, a concept that's pivotal for handling asynchronous tasks. This makes Rust particularly adept for blockchain applications, where multiple network-level requests are made and may not resolve immediately, especially in multi-node environments. A future in Rust effectively encapsulates a pending result, which becomes available once the underlying operation completes. This delayed resolution characteristic of futures is what enables Rust to manage complex, asynchronous tasks efficiently, making it a powerful tool in scenarios such as blockchain, where delayed responses are commonplace.

Errors are common in programming and different languages deal with them in different ways. However, Rust makes it extremely easy for developers to locate and debug errors as it mandates that you recognize the potential for errors and implement necessary measures before code compilation. This demand enhances the resilience of your program by guaranteeing the identification and proper management of errors before deploying your code in a production environment.

Rust is a **memory-safe** language that employs a compiler to track the ownership of values that can be used once and a borrow checker that manages how data is used without relying on traditional garbage collection techniques.

These factors make Rust a perfect candidate for working with blockchains.

In the next section, we will delve deeper into the details of how Rust gives blockchains an edge.

Rust's advantage for blockchains

Blockchains and blockchain-related technologies that use Rust have an edge over others and this section is dedicated to exploring this aspect. Let's learn about how these technologies benefit from using Rust.

Blockchains that use Rust

Some of today's most popular blockchains, such as Solana, NEAR, and Polkadot, use Rust primarily. Polkadot even has a framework called Substrate that can be used to build new blockchains (we have a chapter dedicated to it – that is, *Chapter 10, Hands-On with Substrate,* and this is the framework that was used to build Polkadot itself.

Many new, highly innovative blockchains such as Aptos and Sui also use Rust. In *Chapter 1, Blockchains with Rust*, we learned how Aptos uses parallel processing at the Layer 1 level to make the network extremely scalable.

Hyperledger's Sawtooth is an open source, enterprise-ready blockchain solution for building, deploying, and running distributed ledgers, and this project is also built entirely on the Rust programming language.

Elrond has a WebAssembly virtual machine that utilizes Elrond's web assembly framework and is also based on Rust.

The use of Rust is growing rapidly among blockchain projects and the Web3 community at large and in the coming years, we will see an exponential increase here.

In the following subsections, we will learn about some technologies that make working with blockchains a bit easier. The best part is that these technologies are based on Rust.

Foundry for Ethereum

Ethereum, one of the most popular blockchains, requires smart contracts to be written in the Solidity programming language. But to build, compile, test, and deploy these contracts based on different environments, we usually use a framework such as Hardhat, Truffle, or Web3JS, which not only provides us with the packages but also with a proper project structure to build production-level applications in.

Now, all the frameworks we mentioned are in JavaScript/Node.js. However, Rust engineers also have an option, and that is **Foundry** – a fast, portable, and modular toolkit for Ethereum but one that uses Rust. You get all the benefits of using Rust, such as cargo for managing dependencies and compiling the contracts.

Since we're learning about Rust specifically from the context of it being relevant for blockchains, it is important to learn about Foundry, which makes it simple to work with one of the most popular blockchains.

As part of Foundry, you get three packages – *Forge*, an Ethereum testing framework, *Cast*, a tool for interacting with smart contracts and on-chain EVM data, and *Anvil*, a local Ethereum node, just like Ganache or the Hardhat network.

So, Foundry is a comprehensive, capable toolset for working with Ethereum but solely using Rust.

The Fe, Move, and ink! languages

New programming languages are being built with Rust that are specifically designed for building smart contracts. These languages provide us with the benefit of Rust but at the same time restrict the scope to reduce the size of the executable and also reduce the learning curve in comparison to Rust.

Fe is an alternative to the popular Solidity language for building smart contracts for the Ethereum blockchain and all other chains that use the **Ethereum Virtual Machine** (**EVM**). Fe is simple and the syntax is inspired by Python. It is built with Rust and introduces concepts such as constant generics, which let a user write clean code without sacrificing compile-time guarantees (we learned about zero-cost abstractions previously; this is the same concept).

Fe uses the same intermediate language (Yul) as Solidity, which means it's not just a great choice for Ethereum *mainnet* but also many Layer 2 solutions.

Move is a new programming language that was created by the Diem Association, backed by Meta, and built with Rust. It's built for applications such as blockchains, where safety and correctness are paramount. It is an executable bytecode language that's designed to provide safe and verifiable transaction-oriented computation.

ink! is a programming language developed by Parity Technologies, the parent company of Polkadot and Substrate, for writing smart contracts on blockchain platforms built using the Substrate framework. This language is specifically tailored to enhance the development of smart contracts, leveraging Substrate's capabilities to offer a robust environment for blockchain applications. **ink!** is not to be confused with the INK language, which is written in Go. ink! is completely written in Rust and all the contracts written with ink! are compiled into WebAssembly.

In this section, we learned about the innovations that are happening in the Rust ecosystem, specifically in the context of blockchains. With time, Rust will only be more important and it'll be critical to learn Rust to build on blockchains.

Interesting blockchain projects built with Rust

We already know about various blockchains that are built with Rust (Solana, Polkadot, NEAR, Aptos, and Sui), but apart from these blockchains, there are many interesting projects in the blockchain space that use Rust. Here are a few:

- **Comit** is a project that enables cross-blockchain interoperability between multiple popular chains such as Ethereum, Monero, and Bitcoin. It doesn't introduce a blockchain of its own to provide this functionality, so it is in the blockchain space, without being a blockchain itself.

- **Bonfida** is a token vesting, open source program that enables you to declare timelines for your tokens to get vested/distributed among your founding team, developers, and early users.

- **Astar** is a multi-chain smart contract project built with Rust. It empowers developers to write smart contracts for a single chain, at which point the smart contract can be used across multiple blockchains and virtual machines. While Astar and Comit are both trying to solve the interoperability problem, they're both doing it differently and using a different approach.

- **The Graph** is another highly innovative project. It is a protocol for building decentralized applications on Ethereum and IPFS but using GraphQL. It's open source and built primarily with Rust.

The list of new blockchain projects is increasing every day and more developers are choosing to use Rust to build innovative projects in the blockchain space, especially projects that enable the development of blockchains or smart contracts.

Advantages of Rust-based languages compared to Solidity

In the past few sections, we learned about Fe, Move, and ink! – all languages that are used for smart contract development, but all of them are built using Rust. The other very popular programming language for creating smart contracts is Solidity and a common question that comes up is how Solidity compares to Rust-based languages.

The main benefits Solidity provides are that it's ultra-light and is very simple to learn, especially for people coming from a JavaScript background. This is the reason why some developers feel that Solidity is better for smart contract development usage compared to Rust since Rust is quite extensive and heavy, has longer build times, and also has quite a steep learning curve.

But the main difference to understand here is that Rust is the preferred language for blockchain or protocol development and not just smart contract development, meaning that Rust has a much wider and more complex use case. However, let's also address the usage of Rust specifically for smart contract development, where Solidity is considered a great language to use.

For smart contract development, instead of using Rust, which is quite extensive, there are options such as Move, ink!, and Fe, which are based on Rust, meaning that they provide most of the benefits of Rust but provide this functionality in a simple and light package that's also quite easy to learn and understand. This completely bridges the gap between the benefits that Solidity provides over Rust.

There are also quite a few limitations with Solidity that Rust-based languages overcome easily. For one, smart contracts that are built with Solidity usually have way more vulnerabilities than the ones built with Rust.

Since Rust is super-optimized, smart contracts end up consuming fewer gas fees compared to those built with Solidity.

Let's say we wanted to represent a Bbock in Rust. Here, we can use a feature called **struct**, which enables us to define a custom data type by creating a collection of data types that Rust already understands.

The following code shows a block from a blockchain that uses a Rust `struct`:

```
pub struct Block {
    pub id: u64,
    pub hash: String,
    pub previous_hash: String,
    pub timestamp: i64,
    pub txn_data: String,
    pub nonce: u64,
}
```

In the preceding code snippet, hash and previous_hash are represented as strings. timestamp is a **64-bit signed integer (i64)** and nonce is a **64-bit unsigned integer (u64)**. txn_data is simply the data of transactions that the block will hold. We have used String to represent this, but in the real world, it's going to be a data structure such as an **array** or **vector**. We will implement that when we build the blockchain later in this book.

Let's observe how a blockchain can be represented:

```
pub struct Blockchain {
pub blocks: Vec<Block>,
}
```

Notice that by using structs, we can also represent an entire blockchain, which is simply a collection of multiple blocks. We have represented a collection of blocks by using a Rust feature called Vec, which enables us to store multiple structs. So, in short, a vector is a collection of structs.

> **Note**
>
> In a blockchain struct, there's no need for logic to *connect* the blocks, and this simplistic implementation throws off inexperienced engineers. The fact is that the *connection* logic is inherently present inside the *block* itself, where we store the hash value of the previous block in the previous_hash field.

Now that we have a basic idea of the Rust language, it's time to start learning the concepts that will help us build our blockchain in the next few chapters.

Learning basic Rust concepts

While Rust is quite an extensive language and to cover it completely requires a book of its own, you are advised to supplement your learning with a dedicated Rust book. You can find some examples at https://doc.rust-lang.org/book/. However, the most important concepts that we will often require when working with blockchains can be covered quickly. This is what we will attempt to do here.

First, let's talk about variables and constants. These are the basic building blocks of any programming language.

Variables and constants

Variables are crucial to Rust as they are values that may change multiple times throughout the lifetime of the program. As can be seen in the following example, defining variables in Rust is extremely simple:

```
fn main() {
    let x = 5;
```

```
    println!("The value of is:{x}");
}
```

In the preceding code, we assign a value of 5 to x, where x is the variable, after which we print out the value of x. The code is enclosed in the *main function*, which is the entry point in all Rust programs.

However, by default, as a safety check and to avoid unpredicted behavior, all variables in Rust are immutable (you reassign something to them), so the following example won't work:

```
fn main() {
    let x = 5;
    println!("The value of is:{x}");
    let x = 6;
    println!("The value of is:{x}");
}
```

The preceding program will give us an error. However, there is a way to make variables *mutable*, and that is by adding the mut keyword:

```
fn main() {
    let mut x = 5;
    println!("The value of is:{x}");
    x = 6;
    println!("The value of is:{x}");
}
```

Let's quickly talk about constants. They're similar to immutable variables but with some differences – you can't (obviously) use the mut keyword with them, which means they aren't immutable by default – instead, they are always immutable. Next, you can use the const keyword instead of let to define constants and you must always mention the type. The last difference is that you can't set the value of a constant to a value that is generated from a dynamic computation; it will always be a constant expression. The following code block shows an example of a constant declaration:

```
const HOURS_IN_A_DAY = 24;
```

An interesting property that needs to be pointed out here is *overshadowing*. Here, if we use the let keyword to assign a new value to x, we will not get an error. Let's look at an example:

```
fn main() {
    let mut x = 5;
    println!("The value of is:{x}");
    let x = 6;
    println!("The value of is:{x}");
}
```

In this example, we didn't use the mut keyword again when reassigning the value of x to 6 as this will overwrite the value of x and the new value will become 6. Please note that when you run this code example, you will receive a warning regarding re-assignment, but there will be no error here.

Data types

As we learned in the *Statically typed* section, Rust is a statically typed language, meaning it has a requirement that it needs to be aware of the type of every single variable at compile time. Due to this, it is important to take a look at the various data types in Rust.

Let's get started. In this section, we will look at the *scalar* data types in Rust, namely integers, floating-point numbers, Booleans, and characters. In the *Tuples and arrays section*, we will learn about *compound* data types – tuples and arrays.

Let's start with **integers** – they can either be signed or unsigned. **Signed integers** are represented with an *i* – for example, i16 and i32, while **unsigned integers** are represented with a *u* – for example, u16, and u32. Here, *signed* and *unsigned* refer to the possibility of the occurrence of a negative number and the digits 16 and 32 represent the number of bits of space that the variable is going to take. Additionally, Rust includes **isize** and **usize** types, which are architecture-dependent integer types. The **isize** and **usize** types are primarily used for indexing collections and interfacing with system calls, with their size varying based on the underlying machine architecture.

The most commonly used bits are 8, 16, 32, 64, and 128 and the default integer types that Rust considers are i32 and u32.

The next scalar data type is **floating-point numbers**, which (unlike integers) are numbers that can store decimal points. All floating-point types are signed (again, this is different from integers). The two sub-types that are present are f32 and f64, and they represent 32-bit and 64-bit, respectively. Unlike integers, the default here is 64. While f32 has single precision, f64 has double the precision but similar speed.

The following code block shows floating-point numbers in action:

```
fn main() {
    let x = 2.0;
    let y: f32 = 3.0;
}
```

In the preceding example, in the first line, since we don't mention the number of bits, Rust selects 64-bit by default. In the second example, 32-bit has been mentioned clearly by us.

The next data type is Boolean and it has two possible values – true and false. Let's look at an example:

```
fn main() {
    let t = true;
    let f: bool = false;
}
```

In the preceding example, we can see that we can either assign the Boolean value directly to the variable or we can mention `bool` (as can be observed on the second line), which is how we indicate the Boolean type to Rust.

The last scalar type is `char` and we can use it like this:

```
fn main() {
    let t: char = 'z';
    let f: char = 😎;
}
```

As we can see, `char` can represent a lot more than just **American Standard Code for Information Interchange (ASCII)** characters. Now, it's time to move on to *compound* data types.

Tuples and arrays

The two compound data types present in Rust are tuples and arrays. First, let's take a look at tuples. These are comma-separated lists of values inside parentheses where each value has a type.

There are two important things to remember with tuples – firstly, each value in the tuple can have a different type and they don't have to be the same, and secondly, tuples have a fixed length and once this length has been declared, tuples cannot grow or shrink in size. An important additional point is that tuples are sum data types, which means their total size is the aggregate of the sizes of all contained elements, along with any necessary padding. To better understand tuples, let's look at an example:

```
fn main() {
    let tup = (500, 6.4, 1);
    let (x, y, z) = tup;
    println!("the value of y is: {y}");
    let x:(i32, f64, u8) = (500, 6.4, 1);
    let five_hundred = x.0;
}
```

There are five lines in the preceding example. We will break these down to expand our understanding.

The *first line* defines a tuple called `tup` and assigns a list of three comma-separated values to it so that we can access individual values from `tup`. We can *deconstruct* it, as we have done in the *second line*, to print the value of `y` in the *third line*.

In the *fourth line*, we can be more explicit in defining our tuple and can mention the type of every single value to be more specific and for more control. The *fifth line* shows another way to access individual values from the tuple. So, with this extensive example, not only have we learned how to assign values to a tuple but also how to access those values. One thing to note is that in the preceding case, `tup` will have a padding size of i32 + f64 + u8.

Now, let's look at the second compound data type in Rust: **array**. Now, arrays are also a collection of values, just like tuples, but with a major difference – all the values in an array must be of the same type. Just like tuples, arrays also have a fixed length and are useful when you want to ensure that you always have a fixed number of elements. Let's look at an example to explore arrays:

```
fn main() {
    let a = [1, 2, 3, 4, 5];
    let months = ["jan", "feb", "mar"];
    let b: [i32; 5] = [1, 2, 3, 4, 5];
    let z = [3; 5];
    let first = a[0];
}
```

In the preceding example, in the *first line*, we can see how we can assign an array to a variable. The *second line* also demonstrates the same but all the values being assigned are characters and these two variables (a and months) make it clear that all the values need to be of a specific type. In the *third line*, there's a more explicit definition where we not only specifically mention the type of the array but also mention the length.

The *fourth line* is slightly different and if you were to print the value of z, you would get [3, 3, 3, 3, 3] as the output. This is because in [3; 5], 3 denotes the value that will be stored in the array and 5 represents the number of times it will be stored or the length of the array. So, this gives us a great way to define an array that has repeat values.

The *last line* demonstrates how to access the values of an array with the help of the index, where the index starts with 0 and goes all the way up to n-1, where n is the length of the array. So, the a array starts with 0 for the first value and goes to 4 for the last value.

Numeric operations

All the basic mathematical operations are supported in Rust – for example, addition, subtraction, multiplication, division, and remainder. Let's look at an example to understand this:

```
fn main() {
    let sum = 15 + 2;
    let difference = 15.3 - 2.2;
    let multiplication = 2 * 20;
    let division = 20 / 2;
    let remainder = 21 %2 ;
}
```

The mathematical operators that are used in Rust, such as / for division and % for remainder division, are standard and are the same as any other programming language. The preceding example is quite straightforward and self-explanatory.

In Rust, the memory layout is an essential concept and mainly comprises the stack and the heap, along with **virtual tables (v-tables)** for polymorphism. First, let's take a look at the stack.

Stack

The stack in Rust is a region of memory that's used for static memory allocation. It operates in a very organized last-in, first-out manner. Variables stored on the stack have fixed sizes known at compile time. This makes stack operations incredibly fast as it's just about moving the stack pointer up and down:

```
fn main() {
    let x = 5;      // Integer stored on the stack
    let y = true;   // Boolean stored on the stack
    let z = 'a';    // Character stored on the stack
}
```

In the preceding example, x, y, and z are all variables with sizes known at compile time. Rust allocates space for these variables directly on the stack. Each of these variables is pushed onto the stack when main() starts, and popped off the stack when main() completes. The efficiency of the stack comes from its predictability and the simplicity of the push/pop operations. Now, let's move on to the heap.

Heap

The heap is crucial for dynamic memory allocation in Rust. It is where variables or data structures whose size might change or is unknown at compile time are stored. Since accessing the heap involves following pointers and more complex management, it's slower compared to stack operations:

```
fn main() {
    let mut s = String::from("hello"); // String stored on the heap
    s.push_str(", world!");            // Modifying the string
}
```

In the preceding code, s is a String type, which is mutable and can change size. Initially, Rust allocates memory on the heap for "hello". When we modify s using push_str, it might need more space than initially allocated. The heap allows this flexibility, but it requires Rust to manage the memory, keep track of its size, and potentially move it if more space is needed. Next, we'll look at v-tables.

V-tables

V-tables enable Rust to support dynamic dispatch, particularly with trait objects. A v-table (or virtual method table) is a mechanism that's used in object-oriented programming for method resolution at runtime:

```
trait Animal {
    fn speak(&self);
```

```
    }

    struct Dog;
    struct Cat;

    impl Animal for Dog {
        fn speak(&self) {
            println!("Dog says Bark!");
        }
    }

    impl Animal for Cat {
        fn speak(&self) {
            println!("Cat says Meow!");
        }
    }

    fn make_sound(animal: &dyn Animal) {
        animal.speak();
    }

    fn main() {
        let dog = Dog;
        let cat = Cat;
        make_sound(&dog);
        make_sound(&cat);
    }
```

In this expanded example, make_sound is a function that takes a trait object, &dyn Animal. We have two structs, Dog and Cat, each implementing the Animal trait. When make_sound(&dog) and make_sound(&cat) are called, Rust uses the v-table of each object (Dog and Cat) to look up and call the appropriate speak method. The v-table is essentially a lookup table where Rust can find the correct method implementations for a trait object at runtime, allowing for polymorphism.

Slices

Slices are defined as follows:

"A slice is a pointer to a block of memory and can be used to access portions of data stored in contiguous memory blocks."

To learn more, please refer to https://www.tutorialspoint.com/rust/rust_slices. htm. This definition might be slightly confusing to us right now as we haven't covered the concepts of accessing memory and pointers yet. So, to understand slices with ease, let's break down the complexity slightly.

Slices enable us to refer to a part of a string or an array. Let's look at two separate examples that will help us understand this statement:

```
fn main() {
    let n1 = "example".to_string();
    let c1 = &n1[4..7];
}
```

In the preceding example, if you were to print the value of c1, you would get ple, because c1 only refers to a part of the n1 string.

Let's see how this works – the values in the square brackets, [4..7], indicate that we want to refer to the values starting from the fifth value, which is m, all the way to the last value, e. However, the fifth value itself is not included in this because that's where the counting starts and hence pleis is taken from n1 into c1.

Something is interesting on *line 2* – the presence of &. This is the reference operator and it's used to create a copy of the desired values into c1 without affecting the original value of n1. We will look at this in more detail in the *Ownership and borrowing* section. You may have also noticed to_String() in the *first line*; we will discuss this in the next section.

Slices are not just useful for strings, but for arrays of numbers as well, so let's learn about this via an example:

```
fn main() {
    let arr = [5, 7, 9, 11, 13];
    let slice = &arr[1..3];
    assert_eq!(slice, &[7, 9]);
}
```

In the preceding example, we take an array called arr and then a slice that selects values starting from the first value and until the third value, which simply means 7 and 9. This is why in the *last line*, we assert whether the slice that we have created by selecting a value from arr is equal to the slice containing 7 and 9.

Strings

Working with strings is straightforward in Rust, so it's important to know the difference between the String type and **string literals**. String literals enable us to hard-code some text into our program, as shown in the following code block:

```
fn main() {
    let s = "hello";
}
```

In the preceding example, s is a variable that is equal to "hello". Now, we can't perform any operations on this string literal because, in Rust, string literals are immutable (this is done for better stability). This is why we have the String type:

```
fn main() {
    let mut hello = String::from("hello");
    hello.push('w');
    hello.push_str("world!");
}
```

In the preceding example, note that we can create a mutable variable called "hello" that will be of the String type from a literal string, "hello", with the help of String::from. String is part of the std library. Here, we get access to methods such as push, which enables us to append a char type to the String type ("hello", in this case). The push_str method enables us to append a String type ("world!") to our existing string.

The topic of strings is incomplete without discussing 'to.string()'. Here's an example:

```
fn main() {
    let i = 5;
    let five = String::from("5");
    assert_eq!(five, i.to_string());
}
```

In the preceding example, we take a variable, i, which has a value of 5. This is a number i64. After that, we take a variable, five, which converts the number 5 into a string.

In the last line, we use the assert function to try and compare the values of five and i. We know that one of them is a string and the other is a number, so we use the to_string() method to convert the value of i into a string. Now, when we compare their values, they will be equal.

Enums

Enums in Rust provide us with a way of enlisting different values for a particular value. For example, if you wanted to enlist different types of proxy servers, you would list forward proxy and reverse proxy as the two different values that can be used. For instance, let's consider an example where we define different types of cache strategies.

First, let's redefine our CacheType enum to represent two types of cache strategies – **least recently used (LRU)** and **most recently used (MRU)**:

```
enum CacheType {
    LRU,
    MRU,
}
```

Now, we can create variables of this enum type:

```
let lru_cache = CacheType::LRU;
let mru_cache = CacheType::MRU;
```

Here, `lru_cache` and `mru_cache` are instances of `CacheType`. We use the `::` syntax to specify which variant of the enum we want to create.

Next, let's consider a struct named `Cache`. This struct will have a field that uses our `CacheType` enum. Note that we avoid using "type" as a field name since it's a reserved keyword in Rust. Instead, we'll use `cache_type`:

```
struct Cache {
    level: String,
    cache_type: CacheType,
}
```

In this example, the `Cache` struct has two fields: `level`, which is a string, and `cache_type`, which is of the `CacheType` type. This demonstrates how enums can be integrated into structs, offering a structured and clear way to define data.

By using enums, we set a clear, defined set of values for a variable type, reducing errors and misunderstandings in our code. This is especially helpful for collaborative programming, ensuring consistency and clarity. The use of structs, which we will explore in more detail later, further enhances this by allowing us to create complex data types that incorporate these enums.

Enums in Rust are not just for listing values; they offer a wide range of functionalities. Beyond the basic use demonstrated earlier, enums can serve several other purposes:

- **C-type enums**: In Rust, enums can behave like C-type enums, where each variant is automatically assigned an integer value starting from 0 or a predefined value if specified:

  ```
  enum StatusCode {
      Ok = 200,
      BadRequest = 400,
      NotFound = 404,
  }
  ```

  Here, `StatusCode::Ok` has a value of 200, `BadRequest` has a value of 400, and so on.

- **Enums with values**: Rust enums can also hold data. This feature allows more complex data structures than simple named values:

  ```
  enum CacheStrategy {
      LRU(String),
      MRU(i32),
  }
  ```

In this example, LRU holds a `String` type, and MRU holds an `i32` type. This makes enums incredibly powerful for diverse data representations.

- **Enum size**: An important characteristic of enums in Rust is that the size of an enum type is determined by the largest variant it can hold. This is crucial for understanding memory allocation when using enums.

By understanding these advanced aspects of enums, we can appreciate their versatility and the powerful role they play in Rust's type system. Enums go beyond simple value substitution, allowing for complex data representations and controlled memory usage. With this knowledge, we can use enums to create more efficient and expressive programs in Rust.

Now that we have a firm grip on the basic concepts of Rust, let's learn some intermediate concepts that will help us in the next chapters when we build actual projects.

Exploring intermediate Rust concepts

In the previous section, we understood a lot of the foundational concepts of Rust. This will help us in reading and deciphering larger programs that we shall encounter in the later chapters of this book. Now, it's time to build upon the knowledge that we've acquired to start understanding slightly more advanced Rust concepts so that we gain a stronger grip on the concepts and deploy them in real-world scenarios.

Control flow

All programming languages have a way to execute different pieces of code that are dependent on meeting a condition. This execution happens with the help of branches, which are usually defined with the help of `if`, `else`, and `else if`. In Rust, we have similar concepts. So, let's learn about these with the help of some examples:

```
fn main() {
    let i = 5;
    if i > 3 {
        println!("condition met, i is greater than 3");
    } else {
        println!("condition was not met");
    }
}
```

In the preceding example, we start with a variable, `i`, which is assigned a value of 5. Then, we compare its value to 3 to see if it's greater than 3. Since 5 is greater than 3, the first condition with `if` is met and we print `"condition met, i is greater than 3"`. However, if the value of `i` was 2, then the `if` condition wouldn't have been met and the program would enter the second branch for the `else` condition, so we would print `"condition was not met"`.

Let's explore this further with another example:

```rust
fn main() {
    let a = 10;
    if a % 4 == 0 {
        println!("a is divisible by 4");
    } else if a % 3 == 0 {
        println!("a is divisible by 3");
    } else if a % 2 == 0 {
        println!("a is divisible by 2");
    } else {
        println!("a is not divisible by 4,3, or 2");
    }
}
```

The preceding example demonstrates multiple branches with `if` and `else if`, where we go through multiple conditions to check the different numbers that a (with an assigned value of 10) is divisible by. Now, since 10 is divisible by 2 and the remainder will be 0, the second `else if` condition is met and we print `"a is divisible by 2"`.

While loops

In Rust, a `while` loop allows for repeated execution of a block of code, so long as a specified condition is true. It's a useful tool for implementing situations where the number of iterations is not known in advance but is determined by a condition that's evaluated at each iteration.

Here's an example of a `while` loop in Rust:

```rust
fn main() {
    let mut number = 1;

    while number <= 5 {
        println!("Number: {}", number);
        number += 1;
    }

    println!("Loop exited. Number is now {}", number);
}
```

In this example, the `while` loop continues to run as long as number is less than or equal to 5. Inside the loop, we print the current value of number and then increment it by 1. Once number exceeds 5, the loop exits, and the program continues with any code that follows. This demonstrates how `while` loops can be used for repeating actions until a certain condition changes, making them a versatile tool for various programming scenarios.

Understanding `while` loops is a stepping stone to grasping more advanced concepts in Rust. This knowledge will be particularly useful as we start to tackle more complex programming challenges and apply our skills to real-world scenarios.

Functions

Functions are the core building block of any programming language and the same is true for Rust. So far, we've been working with `fn main()`. This is the main function inside which the starter code of a program is situated and is where the control flow begins. However, there are usually many more functions inside a program and they can be named based on convenience and ease of remembering. Functions help us by breaking down a large piece of code into smaller contextual blocks that make the code easy to read, debug, and execute. They also group repeatable chunks of code so that instead of copying the code that repeats, we have a function that can be called repeatedly. This cements the **Don't Repeat Yourself (DRY)** principle.

Let's explore this with an example:

```
fn main() {
    let a = plus_ten(1);
    println!("the value of a is: {a}");
}
fn plus_ten(a: i32) -> i32 {
    a+10
}
```

In the preceding example, we can see a function called `ten`. We can define a function in Rust by using the `fn` keyword and then round brackets, `( )`, after the name of the function. These round brackets accept *parameters* into the function. In our case, the parameter is `a` and the type is `i32`. Now, after `->`, we see `i32` mentioned once more. This is simply the return type of the function, meaning that the function returns a value that's of the `i32` type. Inside this function, something very simple is happening – `10` is being added to the value that this function receives that's being returned. This is why it has the apt name of `plus_ten`.

An interesting observation here is that we haven't used the `return` statement in the `plus_ten` function. This is because Rust understands the difference between *statements* and *expressions*. Notice that after `a+10`, we haven't used `;` (a semi-colon) because if we use one, Rust treats it as a statement. In Rust, statements are actions that are performed without returning a value, whereas expressions do return a value. In the context of functions, an expression at the end of a block (without a semicolon) implicitly returns its value. However, this simplicity mainly applies to functions with a single exit point. For functions with multiple potential return points, explicitly using the `return` keyword becomes necessary. This keyword is crucial for clearly specifying the return value in different parts of the function, ensuring the intended behavior is achieved.

In the `main` function, notice that we're calling the `plus_ten` function but then assigning it to a variable, a – what's happening here? Well, a is going to simply store the value that's returned from `plus_ten`, so the return value from the function is being assigned to a, not the function itself. The benefit of this type of syntax is that you can call functions, get return values, and assign them to variables in your `main` function, at which point you can compare these values or simply print them, just like we're doing here.

Even though we have seen a simple example where we call a function from our `main` function, it is important to note that this knowledge can be extrapolated and we can call other functions from inside regular functions too, not just the `main` function. With this simple example, we have understood one of the most important features of Rust. This will accelerate our learning through the upcoming sections.

Match control flow

A really handy and unique feature in Rust is the ability to **match** values against a series of values or patterns and then execute based on a condition. This feature goes well with the *enum* feature we learned about previously. Combining the two can lead to some great combinations, so let's see an example of the same:

```
enum Web3{ Defi, NFT, Game, Metaverse }
fn number_assign(web3: Web3) -> u8 {
    match web3 {
    Web3::Defi => 1,
    Web3::NFT => 2,
    Web3::Game => 3,
    Web3::Metaverse => 4,
    }
}
fn main() {
    let defi = number_assign(Web3::Defi);
    assert_eq!(1, defi);
}
```

In the preceding example, we have an enum type called `Web3` that has four different values – `Defi`, `NFT`, `Game`, and `Metaverse`. We want to create a function that assigns a particular number to every different value of the `Web3` enum. For this purpose, we have created a function called `numner_assign`. This has a `match` statement that creates multiple branches of code that are executed based on a matching condition, so we assign a value of 1 if we send `Defi` to this function, 2 if we send `NFT` to this function, and so on. The function accepts `web3` as a parameter that is of the `Web3` type. The difference is in the capital `W` in `Web3`; the latter indicates the enum, which is referred to as the variable's `type`. Another useful feature of `match` is the wildcard pattern, `_`, which acts like a "default" case in a C-type switch statement. It's used to match any value not explicitly handled by the other arms of match:

```
fn rate_movie(rating: u8) {
    match rating {
        5 => println!("Excellent"),
        4 => println!("Good"),
        3 => println!("Average"),
        _ => println!("Need improvement"),
    }
}
```

In the rate_movie function, the _ pattern catches any rating that is not 5, 4, or 3, offering a default response. This feature is especially useful in cases where it's impractical to list every possible variant or when the focus is on specific cases while treating the rest uniformly.

This is exactly why enum and match statements are extremely powerful, especially when combined.

Structs

We talked about structs a few sections back. Structs are an extremely powerful feature of Rust as they enable us to create custom data types by combining existing data types that Rust already understands. These custom data types help us to represent complex information and easily interface with databases and external APIs. An important aspect of structs in Rust is that they are sum data types. This means the total memory size of a struct is the sum of the sizes of all its fields, plus any additional padding necessary for memory alignment. Let's look at a few examples:

```
struct Employee {
    name: String,
    assigned_id: u64,
    email: String,
    active: bool,
}
```

In the preceding example, we have defined a struct called Employee, which is a custom data type that is going to help us store specific employee-related data. It is a combination of multiple data types that Rust already understands, such as String, bool, and u64.

The following example makes use of a struct:

```
struct Employee { name: String, assigned_id: u64, email: String,
active: bool, }
fn main () {
let emp1 = Employee {
name: String::from("John Doe") ,
```

```
assigned_id: 1,
email: String::from("jd@company.com") ,
active: true, };
println!("Employee Name: {}", emp1.name);
println!("Assigned ID: {}", emp1.assigned_id);
println!("Email: {}", emp1.email);
println!("Active: {}", emp1.active);
}
```

The preceding example shows us how to initialize a struct. emp1 is a variable that is of the Employee type, where Employee is the struct we defined earlier. Since we're able to use a struct to define types for variables, this is what we meant when we mentioned that structs are used to define custom types in Rust. When we create an instance of Employee, we can access its elements using the dot notation, as shown in the println! statements.

Each value of the struct, such as name, email, and so on, is initialized in the preceding example where initialization simply provides a particular value to create the variable for that struct type.

Vectors

Vectors enable us to create collections of values of the same type. The ideal use case for vectors is to store a list of values of a particular struct type. In the previous section, we learned about structs and looked at an employee struct, where we initialized a single employee using the custom struct type that we defined. In a real-world scenario, we may have to work with a list of various employees. To handle and store this list, we need vectors. Vectors are distinguished from fixed-size arrays by their extensive set of utility functions and the ability to dynamically resize.

For example, consider the Employee struct that we had previously. With vectors, we can not only store a list of Employee instances but also change and read individual elements with ease:

```
struct Employee {
    name: String,
    assigned_id: u64,
    email: String,
    active: bool,
}

fn main() {
    let emp1 = Employee {
        name: String::from("John Doe"),
        assigned_id: 1,
        email: String::from("jd@company.com"),
        active: true,
```

```
    };

    let mut employees = Vec::new();
    employees.push(emp1); // Storing an employee

    // Changing an element
    if let Some(emp) = employees.get_mut(0) {
        emp.email = String::from("johndoe@newcompany.com");
    }

    // Reading an element
    if let Some(emp) = employees.get(0) {
        println!("Employee Name: {}", emp.name);
    }
}
```

In the preceding code, `employees` is a vector that stores `Employee` instances. We use the `push` method to add an employee. To change an element, we use `get_mut` to obtain a mutable reference to an employee, allowing us to modify its fields. For reading elements, `get` provides an immutable reference to an employee, enabling safe access to its data. Let's see another example to understand this further.

A simple vector can be created using `vec!` with the values that will be stored in that vector:

```
let vector = vec![1, 2, 3];
```

Since we sent 1, 2, and 3 to the vector in the preceding example, Rust can infer the type of the vector – in this case, `i32`. If we want to be more specific about the type and need to ensure the creation of a vector of a particular type, we can use the following syntax:

```
let vector: Vec<i32> = Vec::new();
```

In the preceding example, `vector` is a variable that is actually a vector. It's going to store values of the `i32` type. This is represented by `Vec<i32>`. The creation of a new vector is represented by calling the `"Vec::new()"` function.

Vectors have indices and the values inside a vector can be referenced using the index number, like so:

```
let second: &i32 = &vector[1];
```

In the preceding example, we can refer to the second value of our vector by using `&vector[1]` since the index number starts at 0. Something interesting here is `&`. This is known as a *reference*; we will look at this later.

Let's look at another example where we use vectors for what they're meant for – storing collections of data types:

```
struct rectangle {
  w: i8,
  h: i8
}
fn main() {
let mut v = vec![];
v.push(rectangle{w: 3, h: 4});
v.push(rectangle{w: 99, h: 42});
}
```

In the preceding example, we start with a struct called `rectangle` that has a width and height. Then, we initialize an empty, mutable vector, v. In the next two lines, we push two structs into this vector. These two structs are rectangles that we defined earlier. This demonstrates how we can easily store and work with collections or lists of multiple struct objects. For more information, please refer to the official documentation at `https://doc.rust-lang.org/std/vec/struct.Vec.html`.

In the next section, we will dig into some advanced concepts that will take our Rust skills to the next level.

Delving deep into advanced Rust concepts

Now that we have built upon our foundational knowledge of Rust, it's time to expand our bases with some more concepts that will make us even more comfortable with Rust and let us build real-world projects with it.

In the following subsections, we will learn about concepts that unlock some advanced functionality for us, such as hashmaps, ownership, borrowing, crates, modules, and cargo. These help us work smoothly with slightly bigger projects. However, you may end up using these features more often than you expect, so it's important to pay close attention to these subsections.

Hashmaps

Hashmaps in Rust, as in many other programming languages, are collections of key-value pairs. They are especially efficient for scenarios where you need to quickly look up data using keys rather than index values. The efficiency of hashmaps comes from their use of a hashing function, which converts keys into hash codes. These hash codes are then used to determine where to store the key-value pairs in the hashmap's internal array. This mechanism allows for fast data retrieval as it reduces the need to search through the entire data structure; instead, the key's hash code points directly to its associated value.

Let's explore how to use a hashmap in Rust with an example:

```rust
use std::collections::HashMap;

fn main() {
    let mut rgb = HashMap::new();

    rgb.insert(String::from("Blue"), 10);
    rgb.insert(String::from("Green"), 50);
    rgb.insert(String::from("Red"), 100);

    // Querying and using data from the HashMap
    match rgb.get("Blue") {
        Some(&number) => println!("Blue: {}", number),
        None => println!("No value found for Blue"),
    }

    // Iterating over the HashMap
    for (key, value) in &rgb {
        println!("{key}: {value}");
    }
}
```

In the preceding example, we create a mutable `HashMap` named `rgb`. Then, we insert key-value pairs into it, where each key is a color name and the value is an associated number. To query and use data from the hashmap, we use the `get` method. This method retrieves the value associated with a given key – in this case, `"Blue"`. If the key exists, its value is printed; if not, a message indicating the absence of the key is displayed.

The final part of the example demonstrates iterating over `HashMap` using a `for` loop. This loop prints out each key-value pair stored in the `rgb` hashmap.

Hashmaps in Rust are an integral part of efficient data handling, particularly when you need fast access to elements through unique keys. Their underlying implementation makes them a preferred choice over other data structures, such as arrays or vectors, in scenarios where quick lookup is a priority.

Ownership and borrowing

The & character has been in a few examples in previous sections. We discussed that this is called passing values by reference. In this section, we will understand why this is a critical feature provided by Rust.

Rust has an interesting approach to the problem of memory management that many programming languages try and solve with an automated garbage collector. Some expect the programmer to explicitly allocate and free the memory. The approach that's used by Rust involves ownership and a set of rules that the compiler checks for you. This harks back to what we learned about the Rust compiler at the beginning of this chapter – the compiler does a lot of the work for you and prevents you from doing something irrational.

The general rules that define this state that all values in Rust have an owner and that there can only be one owner for each value. The rules also state that when the owner goes out of scope, the value will be dropped.

Let's look at a few examples:

```
fn main() {
        let example = String::from("hello");
  }
```

The preceding example shows a variable called example that implements a string called 'hello'. What we want to demonstrate with this example is that example will go out of scope after the end bracket of the main function. This is where the uniqueness of Rust comes in – as soon as the variable goes out of scope, Rust calls a special function called drop that returns the memory to the allocator.

Let's look at an example of how **ownership** plays a significant role in this process:

```
fn main( ){
    let example = String::from("hello");
    another_function(example);
    println!("{}", example)
}
fn another_function(example String){
    println!("{}", example)
}
```

In the preceding example, the main function has a variable called example that is a string, while the main function has *ownership* of example. However, as soon as we send example off to another_function, this called function gets ownership of example and it has the code to print the value of example. The issue is that since there can only be one owner of a particular value in Rust, the main function does not have ownership of example anymore – this is where most Rust beginners get confused.

This means that once we call another_function with the value of example, the main function loses ownership. Now, when we try to print the value of example in the main function after calling another_function, the Rust compiler isn't going to allow us to do that. This is done to ensure the program is highly stable.

So, if we want to send ownership to another function but also want to use the value in the native function, how do we do this? This is where the concepts of **borrowing** and **reference** come into play.

Let's demonstrate this with an example:

```
fn main() {
    let s1 = String::from("hello");
    let len = calculate_length(&s1);
    println!("The length of '{}' is {}.", s1, len);
}

fn calculate_length(s: &String) -> usize {
    s.len()
}
```

In the preceding example, we pass a value of s1, which is a string to the calculate_length function. However, it's important to note &, which means that we didn't transfer ownership to this function. Instead, ownership resides with main but due to &, only reference to s1 is being borrowed by the calculate_length function.

This means that even if we want to print the value of s1 later in the main function, even after calling the calculate_length function, we can do that and it won't lead to any errors. Previously, we did not have this ability. With these simple examples, we have demonstrated and understood the important concepts of ownership and referencing in Rust.

Crates, modules, and cargo

With Rust, we can achieve modularity by separating the code into *crates*, the smallest compilation units in Rust. What we mean by this is that the Rust compiler (rustc) can compile a crate into a binary (which can be used to run a program) or even a library (which can be used in other projects or other parts of your code). This introduces modularity, reusability, and, to some extent, separation of concerns. This concept also brings in *shareability*, where crates can be shared among programmers across different projects for ease and to avoid building features from scratch.

In Rust, library and package are the same and Cargo is the name of the package management tool that we get with Rust. Both rustc and Cargo come with the standard rustup installation.

The following command creates a crate using Cargo (we will talk about Cargo in a minute):

```
$ cargo new phrases
```

This command generates a simple project for us:

```
.
├── Cargo.toml
└── src
    └── lib.rs

1 directory, 2 files
```

Modules enable us to partition our code within the crate itself. To define modules, we use the mod keyword:

```
mod english {
    mod greetings {
    }

    mod farewells {
    }
}
```

The preceding example shows how we can define a module inside the phrases crate, where english is a module and english has greetings and farewells phrases. Similarly, the crate could contain phrases in many such languages, with each language being a module.

The point of having crates, as discussed earlier, is to be able to use them in different projects. We can import external crates into our program like so:

```
extern crate phrases;
```

The preceding line demonstrates that we're importing the phrases crate that we created previously into our program.

Crates can end up getting large, hence why we had modules. So, if we want to use particular modules from a crate, we can do that as well:

```
extern crate phrases;
use phrases::english::greetings;
use phrases::english::farewells;
```

The preceding example imports the external phrases crate. Previously, we defined the english module in the phrases crate with english having its own greetings and farewells, so we see the second and third lines of the programs taking the help of the use keyword to access particular sub-modules inside the english module.

The Cargo.toml file is a crucial component of any Rust project. It contains metadata about the project and its dependencies. Think of it as a manifest or a blueprint for your project, detailing everything Cargo needs to know to build your crate, from the crate name and version to its dependencies and build configuration.

The standard Rust project structure includes a src directory, where your source code resides. Within src, the lib.rs or main.rs file is where the crate's root module is defined. Additional modules can be created either as submodules in these files or as separate files and directories within the src directory. This structure helps in organizing code into logical units, enhancing readability and maintainability.

To add external dependencies to your project, you must specify them in `Cargo.toml`. Cargo fetches these dependencies from **crates.io** (`https://crates.io/`), Rust's package registry, when you build your project. Installing a crate is as simple as adding a line to your `Cargo.toml` file under `[dependencies]` with the crate name and desired version, then running `cargo build` or `cargo run`.

Summary

In this chapter, we explored why Rust stands out as one of the most favored programming languages, particularly due to its exceptional speed and performance. Moreover, its distinctive attributes, such as type safety, AOT compilation, zero-cost abstractions, and freedom from garbage collection, contribute to its popularity. Following this, we've delved into crucial Rust concepts that lay the foundation for upcoming chapters. These concepts encompass variables, data types, tuples, arrays, slices, control flow, functions, and vectors, all of which will be instrumental in our continued learning journey.

We also covered advanced concepts such as hashmaps, ownership and borrowing, crates, modules, and cargo, all of which help us work with projects.

This was a quick, whirlwind tour of the most important and the most commonly used Rust features that we will be using and reusing throughout the rest of this book, hence why it was important to get this out of the way. However, it must be noted that most of the explanations were quite pithy, so you are encouraged to read more in-depth books dedicated to the Rust programming language itself for a better and deeper understanding of the language.

Having said that, we feel that this chapter provides a good working knowledge of Rust that can be used to build complex projects. As we move on, we may need to use more concepts to build our blockchains. These concepts will be explained in detail as we progress through this book.

In the next chapter, we will use our Rust skills to build blockchains and decentralized applications (dApps).

Part 2:
Building the Blockchain

In this part, we will leverage all that we have learned about blockchains and the Rust concept from the previous part and try and build a fully fledged blockchain.

This part has the following chapters:

3

Building a Custom Blockchain

In this chapter, we will embark on the exciting journey of constructing our very own custom blockchain using Rust. By the end of this chapter, we will have laid the groundwork for our blockchain; in the subsequent chapter, we will finalize its implementation.

By undertaking this compact yet impactful project, we'll gain a profound understanding of the fundamental principles underlying blockchain technology. This hands-on experience will equip us with the knowledge needed to recognize and comprehend the inner workings of other prominent blockchains such as Ethereum, Solana, NEAR, and Polkadot, all of which we'll delve into in future chapters.

This understanding is invaluable when it comes to developing **decentralized applications** (**dApps**) on these blockchains. Armed with a solid foundation, we'll be well-prepared to navigate the intricate world of blockchain technology and make meaningful contributions to the decentralized landscape.

To successfully build our blockchain, we will need to address a series of essential steps in this chapter. These include the following:

- Planning our first blockchain project
- Setting up your local environment
- Getting started with building the blockchain
- Creating the genesis block
- Using helper functions
- Exploring embedded databases

Technical requirements

First, we must get our system ready with the required installations.

The most recommended operating systems to work with Rust are Ubuntu and macOS. You can also install Rust on Windows and work with it but you may run into issues in slightly more advanced and complex programs. If you're on Windows, the most recommended way to go is using **Windows Subsystem for Linux** (**WSL-2**), which enables you to work with Ubuntu inside Windows.

Windows installation

For Windows systems, you can directly download the installer from the official Rust website (https://www.rust-lang.org/tools/install).

To set up Rust on your Windows system, you can follow the instructions provided on the Windows development environment web page (https://learn.microsoft.com/en-us/windows/dev-environment/rust/setup).

Mac installation

For macOS users, the best way to install Rust is by using the Homebrew package manager.

Make sure you have Homebrew installed. If you don't have it, go to https://brew.sh/.

Once you have Homebrew, run the following command:

```
brew install rustup
```

rustup is the toolchain manager that includes the compiler and Cargo's package manager. rustup will enable you to switch between different versions of Rust without having to download other stuff. Once you have rustup, you have to use it to install the Rust compiler (rustc):

```
rustup-init
```

After running this command, either restart the terminal or run the following command to activate your changes in the same terminal:

```
source ~/.bash_profile
```

Optionally, you can run the following command:

```
source ~/.zshrc
```

Once you've done this, you just need to verify your installation:

```
rustc --version
```

Ubuntu installation

For Ubuntu and WSL-2, the installation procedure is the same.

The best way to install Rust is by using the rustup command, like so:

```
curl --proto '=https' --tlsv1.2 https://sh.rustup.rs -sSf | sh
```

You will be prompted with some options; select the **default** installation.

Next, we need to run the following command to add the Rust toolchain directory to the current PATH environment variable:

```
source $HOME/.cargo/env
```

You can verify the installation by requesting Rust's version:

```
rustc --version
```

To work with Rust, you need the `build-essential` package, which will download the **GNU Compiler Collection (gcc)**; without this package, you will get an error. You can install it with the following commands:

```
sudo apt update
sudo apt upgrade
sudo apt install build-essential
```

Here are some commonly used Rust commands that can be very helpful:

- This command will help you update Rust very easily:

  ```
  rustup update
  ```

- This command will enable you to uninstall Rust cleanly:

  ```
  rustup self uninstall
  ```

Now, we can install a code editor to write our programs. In this case, we'll be using **Visual Studio Code (VS Code)**.

VS Code

To be able to write programs that require multiple files and modules, you need a code editor. Fortunately, VS Code has quite a few features that make it one of the best code editors for Rust. So, in this section, we will go ahead and install VS Code. You can find the installation for VS Code for Windows at `https://code.visualstudio.com/docs/setup/windows` and the one for macOS at `https://code.visualstudio.com/docs/setup/mac?ref=hackernoon.com`.

Ubuntu snap

After incorporating VS Code, we need to integrate Rust. Just like Homebrew on Mac, there's the `snap` package manager in Ubuntu, which makes it easy to install packages such as Rust. If you already have `snap` installed, then installing Rust is quite straightforward – you just need to run one command:

```
sudo snap install --classic code
```

Following the integration of VS Code, the procedure seamlessly moves forward to include Rust, resulting in the successful installation of both VS Code and Rust on your Ubuntu machine.

Ubuntu apt

VS Code is available from the official Microsoft APT repositories and can be installed on Ubuntu, including within the WSL-2 environment. Let's learn how to install it:

1. The first step is to update the packages index by running the following command as a user with `sudo` privileges:

    ```
    sudo apt update
    ```

2. Then, you need the `wget` package to be able to hit URLs (we'll need to download the repository from Microsoft's website for this):

    ```
    sudo apt install software-properties-common apt-transport-https
    wget
    ```

3. Now, we must import the Microsoft GPG key using the following `wget` command:

    ```
    wget -q https://packages.microsoft.com/keys/microsoft.asc -O- |
    sudo apt-key add -
    ```

4. Next, we must enable the VS Code repository with the following command:

    ```
    sudo add-apt-repository "deb [arch=amd64] https://packages.
    microsoft.com/repos/vscode stable main"
    ```

5. Once the `apt` repository has been enabled, install the VS Code package:

    ```
    sudo apt install code
    ```

In the future, whenever a new version is released, you can update the VS Code package through your standard desktop software update tool by running the following commands in your terminal:

```
sudo apt update
sudo apt upgrade
```

For users running Ubuntu on WSL-2, these same steps apply. WSL-2 provides a full Linux kernel built into Windows and offers an environment to run a GNU/Linux environment directly on Windows, unmodified, without the overhead of a traditional virtual machine or dual-boot setup. This allows you to install and use software such as VS Code in a Linux environment seamlessly alongside your Windows applications. Installing VS Code on Ubuntu on WSL-2 is especially beneficial for developers who need a Linux-based development environment on their Windows machine. It combines the ease of using Linux tools and workflows with the convenience of Windows. It's important to ensure that WSL-2 is installed and set up on your Windows machine before proceeding with these installation steps.

Once set up, open your Ubuntu terminal in WSL-2 and follow the preceding steps to install VS Code.

macOS

To install VS Code on MacOS, you can start by downloading the VS Code installer from `https://code.visualstudio.com/docs?dv=osx`. Then, follow these steps:

1. Open your browser's download list and locate the downloaded app or archive.
2. If it's an archive, extract the archive's contents. You must double-click for some browsers or select the *magnifying glass* icon if you're using Safari.
3. Drag `Visual Studio Code.app` to the `Applications` folder, making it available in the macOS launchpad.
4. Open `VSCode` from the `Applications` folder by double-clicking the icon.
5. Finally, add `VSCode` to your dock by right-clicking on the icon.

Windows

For Windows, you can directly download the VS Code installer from `https://code.visualstudio.com/docs?dv=win`.

Once you have downloaded it, you can run the installer; it will only take a minute or so to install.

By default, VS Code is installed under `C:\Users\{Username}\AppData\Local\Programs\Microsoft VS Code`.

Alternatively, you can download a ZIP archive, extract it, and run VS Code from there.

rust-analyzer

rust-analyzer will be an important add-on in our development environment since it's an implementation of **Language Server Protocol (LSP)** for the Rust programming language. It provides features such as completion and go-to definition for many code editors.

In our case, we need it to help catch issues and errors while we're writing our code so that we don't end up leaving most of the issues for compilation and can make quick fixes as we go.

rust-analyzer is available as a plugin for VS Code and is one of the reasons we went with VS Code. You can install and activate it in VS Code by going to the **Extensions** tab, searching for `rust-analyzer`, and then installing it – it's just a one-click installation.

Once you've done this, you can take it for a ride by creating a simple Rust program using Cargo.

Cargo

We learned about Cargo in *Chapter 1, Blockchains with Rust*. Cargo is Rust's package manager; it downloads your Rust project's dependencies, compiles your packages, makes distributable packages, and uploads them to `creates.io` (the Rust community's package registry).

Cargo is automatically installed along with the Rust compiler as part of our `rustup` installation, so we don't need to take any special steps to get it set up and we can take it for a test drive. To start a new Rust project with the help of `cargo`, run the following command in a terminal:

```
cargo new rust-blockchain
```

This will create a folder for us where we can build our blockchain program. This also gives us some boilerplate code and structure to get started with, as well as information about our crate. The folder structure for the same will look somewhat like what's shown here:

```
/
rust-blockchain/
├── src/
│   └── main.rs
├── .gitignore
├── Cargo.toml
/
```

You will notice the `src` folder, which consists of a file called `main.rs`. This file is where the main function (`fn main`) will reside and is where the program enters from. Also, notice the `cargo.toml` file, which consists of the list of dependencies that you will need in your program. Initially, you won't see any dependencies here because the boilerplate program that you get here is quite simple and doesn't require any dependencies.

However, as we build, we may need to bring in some external packages to help augment and extend the functionalities of Rust. These packages will be mentioned in the `cargo.toml` file. Based on the list mentioned in this file, all the dependencies are downloaded and stored in the `target` folder.

In situations where you're collaborating with other developers on a project, you may need to share your project with them or they may need to make a copy of the program. In this scenario, they don't need to download the target folder – they can just copy the `cargo.toml` file, along with the source code, and install the program using `cargo update` or `cargo build`. Cargo will automatically go through the list and download all the dependencies on their system.

This saves time, space, and hassle in setting up dependencies when collaborating on a project.

Apart from the dependencies, the `cargo.toml` file also allows you to define the project's name, version, features, and edition.

To run the current boilerplate program that exists in `main.rs`, you can simply build the project first with the following command:

```
cargo build
```

Then, you can run the executable file that is generated from the build process by running this command:

```
cargo run
```

Planning our first blockchain project

Before we start writing the code for our first project, we'll practice doing this by going through a visual planning exercise to help us get a visual understanding of how we will approach the project. This helps in setting a structure for the project and not only acts as a roadmap throughout the development of the project but also enhances our understanding of the core components of the project.

Let's discuss these core components in detail.

Structs

Our program is going to contain multiple structs (we briefly touched upon structs in *Chapter 2, Rust – Necessary Concepts for Building Blockchains*, in the *Advantages of Rust-based languages compared to Solidity* section). **Structs** enable us to create data types using a mixture of data types that Rust already understands. Structs also help us keep our code highly modular, readable, and reusable.

The first thing we will start with is a **block** – the smallest entity of the blockchain that stores important information. In our case, the block will store the `timestamp` value for when the block was created, the hash of the previous `Block` via `previous_block_hash`, the `hash` value for this particular block, the transactions data via `Transaction` (the important data that the block stores), the `nonce` value (number only used once), and the `height` value (number of blocks processing this block in the blockchain). We learned about these concepts in *Chapter 1, Blockchains with Rust*.

Let's take a look at *Figure 3.1*, which illustrates the anatomy of a block:

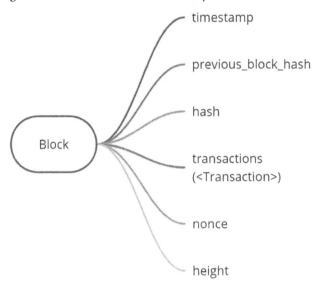

Figure 3.1 – Visualizing the anatomy of a block

A block can have many transactions. The transactions in the block will be stored as a vector of the Transaction struct. Vectors enable us to create collections of structs and we can define a transaction *(Figure 3.2)* as a separate struct that stores the id value for the transaction, vin, which is a vector of **TXInput** (transaction input), and vout, which is a vector of **TXOutput** (transaction output):

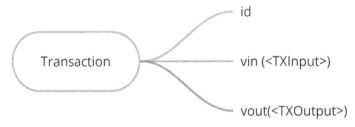

Figure 3.2 – id, vin, and vout

Since vin and vout are vectors of the TXInput and TXOutput structs, let's take a look at what these structs look like. The TXInput struct has txid (the transaction ID), vout – which is again a vector of TXOutput structs. It also has a signature (to sign the transaction data) and a public key (required to access that data).

Let's take a closer look at TXInput in *Figure 3.3*:

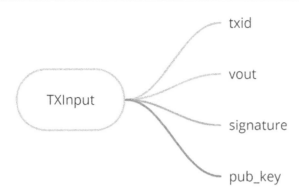

Figure 3.3 – The composition and elements of TXINPUT within a transaction

TXOutput stores the value of the transaction and the public key hash, as depicted in *Figure 3.4*:

Figure 3.4 – The structure and components of TXOUTPUT within a transaction

To uphold the integrity of transaction data within individual blocks, it's essential to utilize three structs: transaction, TXInput, and TXOutput.

A **standard transaction** is comprised of inputs and outputs. Among these, outputs hold paramount significance. These represent indivisible units of cryptocurrency (such as Bitcoin) that are logged on the distributed ledger. The entire network acknowledges these outputs as valid, granting them the ability to be expended. Most transactions work like this, though an exception is a **coinbase transaction**, which does not have input but produces an output – that is, payment to a miner for mining a block.

When a transaction occurs, such as when A sends currency to B, it generates **unspent transaction outputs** (**UTXOs**). These UTXOs are associated with B's wallet address and can be utilized by B. The network of distributed full nodes keeps track of a set of UTXOs, known as the UTXO set. To use the funds that have been received, B has the option to expend one or multiple UTXOs from this UTXO set/pool.

As mentioned in *Chapter 1*, the blockchain is essentially a series of multiple different blocks that are connected. The *connection* between these blocks simply happens because the blocks are maintaining the hash of the previous block. The blockchain stores the tip_hash value, which is simply the hash of the last block, and has a db (database) value, which contains the entire block tree and specifies how blocks are connected, as shown in *Figure 3.5*:

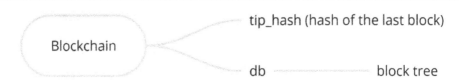

Figure 3.5 – Visualizing block interconnections – tip_hash and block tree in a blockchain

Blockchains have wallets that show the existing balance of the currency with a particular user or account. As shown in *Figure 3.6*, usually, a blockchain wallet stores the private key information with pkcs8 private key information using a format known as **Public-Key Cryptography Standards #8 (PKCS8)**:

Figure 3.6 – Balances and private key storage with PKCS8 encryption

Imagine a user, Alice, who owns a blockchain wallet. This wallet contains a private key, which Alice uses to sign transactions and access her funds. To secure this key, the wallet software uses PKCS8 format. Here's a simplified overview of what happens:

1. **Key generation**: Alice's wallet generates a private key using a cryptographic algorithm such as RSA or ECDSA.

2. **Serialization**: The private key is then serialized into PKCS8 format. This process involves encoding the key into a standard structure that may include additional metadata about the key.

3. **Encryption (optional)**: The PKCS8-formatted key can also be encrypted for added security. This means that even if someone accesses the PKCS8 file, they cannot use the private key without the passphrase.

4. **Storage**: The serialized (and possibly encrypted) PKCS8 private key is stored in the wallet.

This is the key that Alice's wallet will use to sign transactions on her behalf. When Alice needs to make a transaction, her wallet software will use the PKCS8-formatted private key to sign the transaction securely. By leveraging PKCS8, the wallet ensures that the key is stored in a universally recognized and secure format, enhancing the overall security of Alice's assets.

The adoption of PKCS8 in blockchain wallets is an example of how blockchain technology leverages existing, proven cryptographic standards to ensure security. By using PKCS8, wallets can securely store and manage private keys, which are essential for authorizing transactions and maintaining the integrity and security of the user's assets on the blockchain.

Now, let's talk a bit about nodes, which maintain copies of the blockchain we learned about earlier.

Nodes are independent servers that maintain a copy of the blockchain. Some nodes are **full nodes**, which maintain an entire copy of the blockchain. This is really what makes a blockchain decentralized since many nodes are present in a blockchain and even if multiple nodes go down in the network, there are still many nodes to carry the blockchain forward. As shown in *Figure 3.7*, nodes simply contain the address, which is usually the port address if they're on the same server or an IP address if they're on different servers:

Figure 3.7 – Visualizing address inclusion within a network node

Now, let's talk about servers.

As shown in *Figure 3.8*, servers make it easier for us to interact with the blockchain logic via nodes. This means that servers serve the entire blockchain via multiple blocks or the entire block tree:

Figure 3.8 – Visualizing blockchain inclusion within a server

We looked at multiple consensus mechanisms in *Chapter 1*. Now, it's time to code this out. The most fundamental and most widely used consensus mechanism is the **proof of work consensus mechanism** and in our case, proof of work is a struct that stores `block` and `target`. As depicted in *Figure 3.9*, the target is required for mining, where the target is a mathematical result of a formula converted into a hexadecimal number that dictates the mining difficulty:

Figure 3.9 – Visualizing blockchain inclusion within a server

Now, let's learn about the necessary functions for creating a custom blockchain.

Required functions

In this section, we will cover the functions that we require to accompany each struct. This will also dictate how we organize the functions in their respective files – for example, all the functions related to a block will be in a block file. We will discuss the functional planning and file-level planning we'll need to do for our project with the help of diagrams.

Figure 3.10 depicts the elements of a block file:

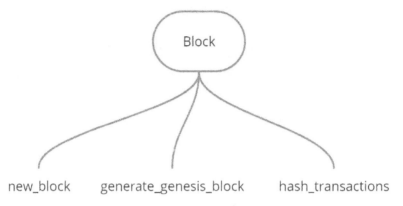

Figure 3.10 – Block file functions

The block file will have a `new_block` function, which is responsible for creating a new block for the chain. The `generate_genesis_block` function helps create the genesis block. The genesis block is the very first block in the blockchain and it requires special care. For example, it won't store any `previous_hash` value; this is why it requires a separate function. The `hash_transactions` function helps us hash the transactions that take place and that are essentially stored in the block.

Let's take a look at the contents of the Blockchain file (*Figure 3.11*).

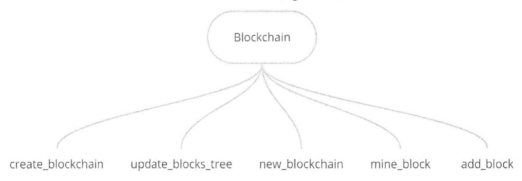

Figure 3.11 – Blockchain file functions

The blockchain file will contain helper functions to create and update the blockchain, such as `create_blockchain`, `update_blocks_tree`, and `new_blockchain`, as well as some additional functions such as `mine_block`, which will mine a new block for the chain, and `add_block`, which will add the newly mined block to the blockchain.

In our program, there will be functions for nodes (*Figure 3.12*) that help us manage and maintain the nodes on which the blockchain exists:

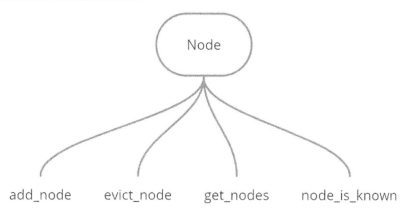

Figure 3.12 – Node file functions

As we know, nodes maintain a copy of the chain and its latest state, and nodes exist in a network where they can communicate with each other to come to a consensus for the next block to be added to the chain.

You can add nodes with the `add_node` function, get all the nodes with the `get_nodes` function, and remove nodes from the network with the `evict node` function.

Figure 3.13 shows the various server functions. The server serves the entire blockchain and each node can be run as a server on a separate port. In the blockchain ecosystem, the terms **node** and **server** often come into play, each serving distinct yet interconnected roles. A node in a blockchain network refers to any computer that participates in the network's activities. These activities include storing data, validating transactions, or contributing to the creation of new blocks. Each node maintains a copy of the blockchain and follows the network's consensus rules.

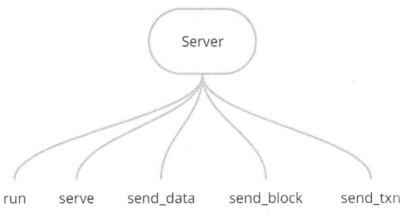

Figure 3.13 – Server functions

On the other hand, a server, in this context, generally refers to a more specialized role. While all servers can be considered nodes, not all nodes are servers. A server typically denotes a node that has additional responsibilities or capabilities, such as serving data or processing requests from other nodes or clients. For example, in a blockchain network, a server node might handle client requests for data, manage connections, or provide an interface for other nodes to interact with the blockchain. When we say a node can **run as a server**, this means that the node has been configured or set up to perform these server-like functions. It's not just passively participating in the network by storing and validating data; it's actively providing services to other nodes or clients. This could involve tasks such as facilitating transactions, offering data query capabilities, or even hosting a user interface for blockchain interaction.

The distinction is important to understand, especially when you're setting up or managing a blockchain network. Each node might have the potential to act as a server, but depending on its configuration and the network's architecture, it might take on a more passive or active role. Running a node as a server on a separate port implies that it is designated to handle specific network functions and communication, often requiring more resources and offering broader capabilities than a regular node:

Our program will have multiple functions for a server, such as `run`, which starts the server, and `serve`, which is the main function that receives requests from the peers (nodes) and then calls the other functions that we have mentioned based on matching conditions. There can be several *matching conditions*, such as data reception, verified blocks, and valid transactions trigger functions for synchronization, network integrity, and consensus among nodes.

The `send_data` function gets the data from the socket, while the `send_block` and `send_tx` functions help advertise the block and the transaction with the rest of the nodes in the network.

Transactions are the core data that the blocks store and they usually take place when money, currency, or value is transferred between accounts, as shown in *Figure 3.14*:

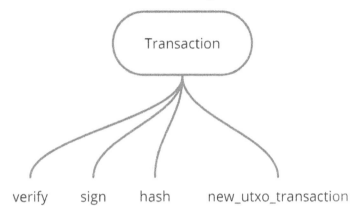

Figure 3.14 – Transaction functions

To ensure the transactions are correct, they need to be verified; for each transaction that takes place, a hash is generated. Also, the account that sends the transaction needs to sign it with its key. This is why we have functions to help support all these functionalities.

As shown in the preceding figure, we have verify for verifying the transactions, sign for signing them, hash to create the hash, and new_utxo_transaction as the unspent transaction output since the receiver of the transaction will have more money in their bank account and will be able to spend this new money they're received.

Wallets can hold value and have addresses associated with them. *Figure 3.15* provides more details:

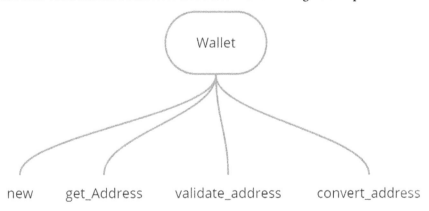

Figure 3.15 – Wallet functions

In our program, we plan to have quite a comprehensive wallet implementation where we will have a new function to create a new wallet, a `get_address` function, which gets the associated wallet address, a `validate_address` function, which checks whether the address is a wallet address or not by using a public key hash, and a `convert_address` function, which takes in a public hash key and gives us a string.

We have organized all the code related to UTXOs into a UTXO set, as shown in *Figure 3.16*:

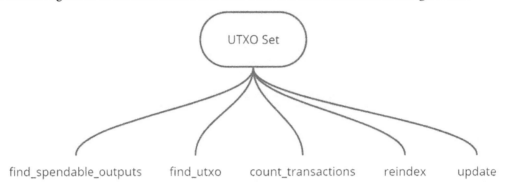

Figure 3.16 – UTXO set functions

In the UXTO set, the `find_utxo` function locates unspent transactions in the complete blockchain tree, `count_transactions` counts the transactions that have taken place, and `find_spendable_outputs` helps us locate spendable UTXOs.

All transactions that take place between accounts on a blockchain make their way to the memory pool (*Figure 3.17*). This is where they are picked and added to a block by the nodes:

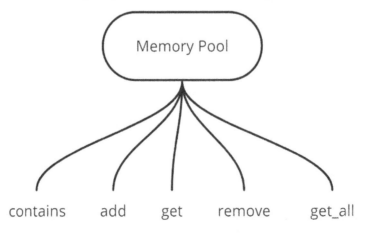

Figure 3.17 – Memory pool functions

All the nodes have access to the memory pool. In our program, we have multiple functions to help work with the memory pool, such as `add` to add transactions to the memory pool. The `get` function will fetch the particular transaction with the provided `tx_id` value from the memory pool, while `get_all` will get all the transactions that are currently in the memory pool. Finally, the `remove` function removes the transactions and the `contains` function checks whether the memory pool contains a particular transaction and returns a Boolean value (`true` or `false`).

With that, we have gone through all the structs and functions that our program will have. This helps create a roadmap of sorts that will guide us on our journey to build a blockchain with a healthy set of features. Even if the functions and their uses don't make sense right now, that's all right – we will have much more clarity when we go through the code.

Now, we need to start coding up our blockchain based on the planning that we've done.

Getting started with building the blockchain

In this section, we'll create our first blockchain project. We will use the Cargo project we created in the previous section as the starting point for our project.

Block

In the upcoming steps, we will guide you through the process of creating a block, elucidating each essential detail along the way. In the `src` folder, create a file called `block.rs` and implement the following struct in it:

```
pub struct Block {
    timestamp: i64,
    pre_block_hash: String,
    hash: String,
    transactions: Vec<Transaction>,
    nonce: i64,
    height: usize,
}
```

We will be following the plan and roadmap that we created earlier, where we did some visual planning, and extend upon this by going through the data types that we have chosen for each of the fields.

Here, `Block` has been represented as a struct where we have the following:

- `timestamp`: An integer value that represents the time when the block was created. It's used to track the chronological order of blocks in the blockchain.

- `pre_block_hash`: A string containing the hash value of the previous block in the blockchain. This creates a link between blocks, ensuring the integrity of the blockchain.

- hash: A string containing the hash value of the current block. This hash is generated based on the data within the current block, including transactions and other information. It's a unique identifier for the block.

- transactions: A vector or collection that holds the various transactions included in the block. Transactions can represent various types of data or actions, depending on the blockchain's purpose (for example, cryptocurrency transactions).

- nonce: An integer called **nonce**, which stands for **number used only once**. It's a value that miners change while mining a block to find a hash that meets specific criteria. It's a crucial part of the proof-of-work consensus algorithm used in many blockchains.

- height: This is a usize value that indicates the position of the current block within the blockchain. It represents the number of blocks that come before the current block. usize is a dynamic size type in Rust that's suitable for representing this variable size.

Once we have the struct to be able to implement a structure for the block, we can take a look at some of the functions that will help us work with blocks. The first step in creating the new block is to initialize a new Block object. The timestamp field is set to current time, and the pre_block_hash, transactions, and height fields are set to the values that were passed in as arguments. The hash and nonce fields are initialized to empty values:

```
pub fn new_block(pre_block_hash: String, transactions: &[Transaction],
height: usize) -> Block {
        let mut block = Block {
                timestamp: crate::current_timestamp(),
                pre_block_hash,
                hash: String::new(),
                transactions: transactions.to_vec(),
                nonce: 0,
                height,
        };
```

This function is used to create a new Block object for a blockchain. It takes in three arguments:

- pre_block_hash: The hash of the previous block in the chain

- transactions: An array of transaction objects that are to be included in the new block

- height: The height of the new block in the blockchain

The function returns a new Block object that has been generated using a proof-of-work algorithm.

The next step is to generate a *proof of work* for the new block using the ProofOfWork struct.

The `ProofOfWork::new_proof_of_work` function takes in a `Block` object and returns a new `ProofOfWork` object:

```
    let pow = ProofOfWork::new_proof_of_work(block.clone());
    let (nonce, hash) = pow.run();
    block.nonce = nonce;
    block.hash = hash;
    return block;
}
```

The `pow.run()` method is then called on the `ProofOfWork` object, which performs the proof-of-work algorithm and returns a tuple containing the nonce and hash values that were found. These values are then assigned to the nonce and hash fields of the `Block` object.

Finally, the function returns the completed `Block` object.

Note that this function uses the proof-of-work struct and transactions struct, so let's define them.

For proof of work, we need to create the `ProofOfWork` struct, which needs to go into the `proof_of_work.rs` file. Here, we have the `block` field, which is the `Block` struct we just created, and `target`, which is a big integer (`BigInt`):

```
pub struct ProofOfWork {
    block: Block,
    target: BigInt,
}
```

The proof-of-work implementation involves a critical step – crafting three essential transaction-related structs. These three structs have to be added to the `transaction.rs` file. The first struct is `Transaction`, where `id` is a vector and `vin` and `vout` are collections of the `TXInput` and `TXOutput` structs, which will also be defined in this file:

```
pub struct Transaction {
    id: Vec<u8>,
    vin: Vec<TXInput>,
    vout: Vec<TXOutput>,
}
```

The `TXInput` struct contains the transaction ID (`txid`), which is a vector, `vout`, which is of the `usize` type, a dynamically determined data type, `signature`, and a public key (`pub_key`) of vectors of `u8`:

```
pub struct TXInput {
    txid: Vec<u8>,
    vout: usize,
    signature: Vec<u8>,
```

```
        pub_key: Vec<u8>,
    }
```

The TXOutput struct has value, which is an integer, and a public key hash (pub_key_hash), which is a vector or collection of u8 values (8-bit unsigned integer types):

```
pub struct TXOutput {
    value: i32,
    pub_key_hash: Vec<u8>,
}
```

In the Block subsection, we saw that the run function from the proof-of-work algorithm was called. First, we'll copy and paste the run function, which plays a crucial role in the mining process in the newly created proof_of_work.rs file. Then, we'll understand what it does:

```
pub fn run(&self) -> (i64, String) {
        let mut nonce = 0;
        let mut hash = Vec::new();
        println!("Mining the block");
        while nonce < MAX_NONCE {
            let data = self.prepare_data(nonce);
            hash = crate::sha256_digest(data.as_slice());
            let hash_int = BigInt::from_bytes_be(Sign::Plus, hash.
as_slice());


            if hash_int.lt(self.target.borrow()) {
                println!("{}", HEXLOWER.encode(hash.as_slice()));
                break;
            } else {
                nonce += 1;
            }
        }
        println!();
        return (nonce, HEXLOWER.encode(hash.as_slice()));
    }
```

This function is part of the proof-of-work algorithm and is used to find a nonce value that produces a hash of the block data that is lower than a specific target value. The function takes no arguments but operates on the ProofOfWork object on which it is called.

The function returns a tuple containing two values: the nonce value that was found, and the hash value that was produced using that nonce value. It works as follows:

1.   The function initiates the nonce variable at 0 and prepares an empty hash vector.

2. It enters a loop, persisting until a valid nonce value or the maximum nonce value is achieved.

3. In each loop iteration, the function does the following:

    I.   It invokes the 'prepare_data' method of the ProofOfWork object, extracting a byte slice depicting block data with the current nonce.

    II. It computes the hash value of this data using the 'sha256_digest' function from the crypto crate (crate:😊).

    III. It converts the hash value into a BigInt value through 'from_bytes_be', treating the hash as big-endian bytes, with 'Sign::Plus' indicating a positive BigInt value.

4. The function checks if the BigInt hash value is less than the ProofOfWork object's target value. If this is true, it prints the hexadecimal hash using 'HEXLOWER.encode' from hex crate and exits the loop. If this is false, the function increments nonce and proceeds to the next loop iteration. After discovering an appropriate nonce, the function returns a tuple comprising the nonce and hash values in hexadecimal form.

Next, we'll craft the very first block in our blockchain – the genesis block. This block sets the stage for our entire blockchain system.

Creating the genesis block

The very first block in a blockchain is called the **genesis block**. While we haven't written any code for the blockchain, we would like to create the first block and get things started. The genesis block doesn't store a previous hash since no block comes before it.

In the block.rs file, we'll begin by implementing functions specific to the Block type. To achieve this, we must create a code block, like so:

```
impl Block{
// Functions related to the Block type can be implemented here
...
}
```

The impl keyword in Rust is used to define a set of methods associated with a specific type – in this case, Block. This block acts as a container for functionality that is directly associated with the block structure, allowing us to define what operations can be performed with or on a block.

Move the new_block function that we created in the *Block* section, where we learned how to create a block, to inside the preceding code block and all the other functions.

Now, we can create a block that will also be placed inside this. First, we will add two more functions inside the code block.

The following two functions are used to serialize and deserialize Block objects to store them in a file, send them over a network, or otherwise persist them outside of memory. They rely on the bincode crate, which is a Rust crate that provides fast and compact binary serialization and deserialization.

The bincode crate is an essential tool for serializing and deserializing Rust data structures efficiently and compactly. To use bincode to handle Block objects, we must add it to our project and import it into our file.

To add bincode to your project, include it in your Cargo.toml file under [dependencies]:

```
[dependencies]
bincode = "1.3.3" # Specify the compatible version
```

After saving the Cargo.toml file, Cargo will automatically download and compile the bincode crate when you build your project.

Now, in the block.rs file where you want to serialize and deserialize Block objects, start by importing the bincode crate:

```
extern crate bincode;
use bincode::{serialize, deserialize};
```

With bincode imported, you can now implement the serialization and deserialization functions:

```
pub fn deserialize(bytes: &[u8]) -> Block {
        bincode::deserialize(bytes).unwrap()
    }
```

The deserialize function takes a slice of bytes and returns a Block object that has been deserialized from the binary data. First, it calls the deserialize method of bincode crate, passing in the slice of bytes. The result of this call is a Result object that contains either the deserialized Block object or an error. The function then unwraps this result and returns the deserialized object:

```
    pub fn serialize(&self) -> Vec<u8> {
        bincode::serialize(self).unwrap().to_vec()
    }
```

The serialize function takes a reference to a Block object and returns a vector of bytes that represents the object in a serialized binary format. First, it calls the serialize method of the bincode crate, passing in a reference to the Block object. The result of this call is a Result object that contains either the serialized bytes or an error. The function then unwraps this result and converts the serialized data into a Vec<u8> struct.

Together, these functions allow Block objects to be easily serialized and deserialized in a binary format that can be stored, sent, or otherwise processed outside of the Rust program that created them.

At the top of the program, include the ProofOfWork crate, since we want to use the run function from it, and the Transaction crate since we want to store transactions in the block. Note that these are the crates we created in our program. Having crates makes programs very modular and easy to import, as shown in the following code snippet:

```
use crate::{ProofOfWork, Transaction};
use serde::{Deserialize, Serialize};
```

We also import serde, which helps us with serialization and deserialization. serde is a popular and widely used library.

The following code represents the content of the updated block.rs file and reflects the changes that have been made by adding the new code:

```
use crate::proof_of_work::ProofOfWork;
use crate::transaction::Transaction;
```

In Rust, serde is a framework for serializing and deserializing data structures efficiently and generically. To use serde in our block.rs file, we need to add it to our project's dependencies and import it into the file.

Here's how we can add serde to the project:

1. Open the Cargo.toml file located at the root of your Rust project.
2. Under the [dependencies] section, add serde and, optionally, specify the version you want to use. If you're using features such as derive, you will also need to enable them:

    ```
    [dependencies]
    serde = { version = "1.0", features = ["derive"] }
    ```

Once you've saved the changes to Cargo.toml, Cargo will handle downloading and compiling the serde crate, along with any specified features.

Next, to import serde in your block.rs file, include the following use statement:

```
use serde::{Deserialize, Serialize};
```

This statement brings the Serialize and Deserialize traits into scope, which are necessary to serialize and deserialize your data structures with serde:

```
impl Block{
pub fn new_block(pre_block_hash: String, transactions: &[Transaction],
height: usize) -> Block {
        let mut block = Block {

                timestamp: crate::current_timestamp(),
                pre_block_hash,
```

```
                hash: String::new(),
                transactions: transactions.to_vec(),
                nonce: 0,
                height,
            };
            let pow = ProofOfWork::new_proof_of_work(block.clone());
            let (nonce, hash) = pow.run();
            block.nonce = nonce;
            block.hash = hash;
            return block;
    }
    pub fn deserialize(bytes: &[u8]) -> Block {
            bincode::deserialize(bytes).unwrap()
    }

    pub fn serialize(&self) -> Vec<u8> {
            bincode::serialize(self).unwrap().to_vec()
    }

}
```

Now, let's delve into crafting some essential helper functions to complement the foundational block structure we've established here.

Using helper functions

Helper functions simplify complex blockchain operations, enhancing code readability, reusability, and maintenance. In Rust-based blockchain development, functions such as get_transactions, get_prev_block_hash, and get_hash abstract away intricacies, enabling modular design and efficient debugging while focusing on high-level logic.

Let's add some functions to the block.rs file that can help us work with blocks:

```
pub fn get_transactions(&self) -> &[Transaction] {
        self.transactions.as_slice()
    }
```

This function helps us to get the list of transactions, but let's break it down and see what's happening here.

This function is defined on a Rust struct and returns a borrowed reference to a slice of Transaction objects:

- pub indicates that this function can be called from outside the struct.

- fn is the keyword that's used to define a function.

- `get_transactions` is the name of the function.
- `(&self)` is the function parameter, which is a reference to the struct instance. `&` indicates that this is a borrowed reference, meaning that the function can read but not modify the struct's data.
- `-> &[Transaction]` is the return type, which is a borrowed reference to a slice of `Transaction` objects.
- `self.transactions.as_slice()` is the implementation of the function. Here, `self.transactions` is a reference to a vector of `Transaction` objects that is a member of the struct.
- The `as_slice()` method is called on this vector to return a borrowed reference to its slice. This slice is then returned from the function.

Therefore, when this function is called on an instance of the struct, it returns a borrowed reference to the slice of `Transaction` objects contained within that instance. Since it is a borrowed reference, the caller cannot modify the contents of the slice directly.

Let's add one more function called `get_pre_block_hash`:

```
pub fn get_pre_block_hash(&self) -> String {
        self.pre_block_hash.clone()
    }
```

This function is defined on a Rust struct and returns a cloned copy of the `pre_block_hash` string. Let's break down the function definition:

- `-> String` is the return type, which is a new `String` instance containing a cloned copy of the `pre_block_hash` string.
- `self.pre_block_hash.clone()` is the implementation of the function. Here, `self.pre_block_hash` is a string member of the struct that holds the hash of the previous block in the blockchain.
- The `clone()` method is called on this string to create a new instance with identical content. This new instance is then returned from the function.

Therefore, when this function is called on an instance of the struct, it returns a new `String` instance that contains a cloned copy of the `pre_block_hash` string held within that instance. The `clone()` method is used to ensure that the returned value is a new, independent copy of the original string, rather than a reference to the original string that could be modified by the caller.

A function similar to `pre_block_hash` is the `get_hash` function; both provide access to hash values. The `get_hash` function provides a reference to a block's hash, offering an efficient way to access the data without taking ownership. This is useful when you only need to read the hash and do not require ownership, such as when you're temporarily inspecting the hash or passing it to functions that don't need to own it. On the other hand, `get_pre_block_hash` clones the previous block's hash, which involves allocating new memory and copying the data. This is necessary when you need to modify the hash or retain it independently of the original block, ensuring that the original data remains unchanged. So, let's go ahead and implement get_hash as well:

```
pub fn get_hash(&self) -> &str {
        self.hash.as_str()
    }
```

This function simply gets the hash for the transaction.

We need three more helper functions that help us return values from the block struct. So, essentially, all the fields of the block will have functions associated to help us isolate them and return them:

```
pub fn get_hash_bytes(&self) -> Vec<u8> {
        self.hash.as_bytes().to_vec()
    }
```

The get_hash_bytes function returns a vector of bytes representing the hash string held within the struct instance. Let's break down the function definition:

- -> Vec<u8> is the return type. It is a new Vec instance containing a copy of the bytes that make up the hash string.
- self.hash.as_bytes().to_vec() is the implementation of the function. Here, self.hash is a string member of the struct that holds the hash of the current block.
- The as_bytes() method is called on this string to obtain a slice of bytes that make up the string's content. This byte slice is then converted into a new Vec instance using the to_vec() method. This new instance is then returned from the function:

```
pub fn get_timestamp(&self) -> i64 {
        self.timestamp
    }
```

The get_timestamp function returns the timestamp value held within the struct instance as an i64 type. Let's break down the function definition:

- -> i64 is the return type, which is the timestamp value as an i64 type.
- self.timestamp is the implementation of the function. Here, self.timestamp is a member of the struct that holds the timestamp of the current block. This value is returned from the function:

```
pub fn get_height(&self) -> usize {
        self.height
    }
```

The get_height function returns the height value held within the struct instance as a usize type. Let's break down the function definition:

- -> usize is the return type, which is the height value as a usize type.

- self.height is the implementation of the function. Here, self.height is a member of the struct that holds the height of the current block.

Then, we have our hash_transactions function. Let's take a look:

```
pub fn hash_transactions(&self) -> Vec<u8> {
        let mut txhashs = vec![];
        for transaction in &self.transactions {
            txhashs.extend(transaction.get_id());
        }
        crate::sha256_digest(txhashs.as_slice())
    }
```

Essentially, this function helps us hash our transactions by taking a collection of transaction IDs and using *SHA-256* to hash them and then return the resulting hash as a vector of bytes. Let's take a closer look:

1. The first line defines the hash_transactions function as a public function that takes a reference to self (which is an instance of some struct that contains a collection of transactions) and returns a vector of bytes (Vec<u8>).

2. In the second line, we initialize an empty vector called txhashs. This will be used to store the hashed transaction IDs.

3. The loop iterates through each transaction in the transactions collection of the self struct, gets the transaction ID using the get_id method, and appends the ID bytes to the txhashs vector using the extend method.

4. The last line calls the sha256_digest function defined in another crate (hence crate:: prefix), passing in the txhashs vector as a slice. This function applies the SHA-256 hashing algorithm to the input data and returns the resulting hash as a vector of bytes.

Finally, we have our generate_genesis_block function:

```
pub fn generate_genesis_block(transaction: &Transaction) -> Block {
        let transactions = vec![transaction.clone()];
        return Block::new_block(String::from("None"), &transactions,
0);
    }
```

As its name suggests, this function generates the genesis block of the blockchain, which is the first block in the chain. Let's take a closer look:

1. The first line defines a function called `generate_genesis_block` that takes a reference to a `Transaction` object as input and returns a `Block` object.

2. The next line creates a new vector called `transactions` that contains a single transaction. This is the transaction that was passed to the function. The `clone` method is called on the transaction to create a new copy of it in memory; this is added to the `transactions` vector.

3. The third line creates a new `Block` object using the `new_block` method of the `Block` struct. The method takes three arguments:

    *   A `String` object that represents the previous block's hash. Since this is the first block, there is no previous block, so a value of `None` is passed.

    *   A reference to the `transactions` vector we created earlier.

    *   An integer value that represents the current block's index. Since this is the first block, the index is set to 0.

4. Finally, the function returns the newly created `Block` object.

Next, we'll look at embedded databases.

Exploring embedded databases

An **embedded database blockchain** employs a built-in database within its nodes, streamlining data storage and retrieval. This approach enhances efficiency and simplifies setup, making it a promising solution for decentralized applications and systems requiring a self-contained blockchain structure.

To store the block tree, we're using an embedded database. We could store these details in a database such as MongoDB, but that would require a significant amount of setup and would pull away our focus from the blockchain logic. So, to keep things simple, we're using an embedded database; this project is for learning purposes and is not intended to be used in production.

However, if you were to use it in production, you could further optimize it, change the database, and then use it in production. But for this book, we will focus on the learning aspect and use an embedded database.

Let's learn more about the embedded database.

We're using the popular **sled database**, which is quite lightweight but still packs a punch in terms of the number of features available. You get features such as the following:

*   Zero-copy reads
*   Write batches

- Merge operators
- Forward and reverse iterators over ranges of items
- CPU-scalable lock-free implementation
- Flash-optimized log-structured storage

These features make it easy for us to quickly implement a solution that's still quick to scale.

One more thing you will notice in the code related to sled will be IVec. You can learn more about IVec here: `https://docs.rs/sled/latest/sled/struct.IVec.html`. IVec is similar to binary and you need this to tell sled how to store your structs in the database. Most structs are easy but for the more complicated structs, we have to explain them to sled. For this reason, we will be making use of `IVec`. To use the `IVec` type from the `sled` crate in our Rust project, we must add `sled` as a dependency. `IVec` is an abstraction over sled's on-disk byte arrays, and it is used extensively within sled's API for data retrieval and manipulation.

Follow these steps to add sled to our project:

1. Open your project's `Cargo.toml` file, which is located at the root of your Rust project directory.
2. In the [dependencies] section, add `sled` with the desired version number. If you do not specify a version, Cargo will automatically use the latest version available. It's often a good practice to specify at least the major version you are targeting to avoid unexpected updates:

   ```
   [dependencies]
   sled = "0.34"
   ```

Once you've added `sled` to your `Cargo.toml` file and saved it, Cargo will handle the rest. The next time you build your project, Cargo will automatically download and compile the `sled` crate, along with its dependencies.

In our `block.rs` file, we must import IVec from `sled`:

```
use sled::IVec;
```

Now that we know about `sled` and `IVec`, let's look at some code related to this that we need to write in the `block.rs` file outside the `impl` block:

```
impl From<Block> for IVec {
    fn from(b: Block) -> Self {
        let bytes = bincode::serialize(&b).unwrap();
        Self::from(bytes)
    }
}
```

This is an implementation of the `From` trait for a `Block` struct and an `IVec` struct. In Rust blockchain development, traits define shared behaviors among types. They enhance reusability, enforce consistent behavior, and promote flexible, polymorphic interactions within the blockchain system. `From trait` allows us to create an instance of one type from an instance of another type.

In this implementation, a `Block` instance is converted into an `IVec` instance. Here's how the implementation works:

- `impl From<Block> for IVec`: This is the implementation of the `From` trait for the `Block` struct and the `IVec` struct. This means that an instance of `Block` can be converted into an instance of `IVec` using this implementation.

- `fn from(b: Block) -> Self { ... }`: This is the implementation of the `from` method for this trait. The method takes a single parameter, b, of the `Block` type and returns an instance of `IVec`.

- `let bytes = bincode::serialize(&b).unwrap();`: This line serializes the `Block` instance into a byte array using the `bincode` crate's `serialize` function. This converts the `Block` instance into a binary format that can be stored or transmitted over a network.

- `Self::from(bytes)`: This line converts the `byte` array into an instance of `IVec` using the from method of the `IVec` struct. `Self` is a special keyword in Rust that refers to the type that is being implemented, which is `From<Block>` for `IVec` in this case. The `from` method of the `IVec` struct takes a byte slice as input and returns an `IVec` instance containing the bytes.

Therefore, when this implementation is used to convert a `Block` instance into an `IVec` instance, it first serializes the `Block` instance into a byte array using the `bincode` crate, then creates a new `IVec` instance from that byte array.

With this, we've successfully crafted our very own custom blockchain.

Summary

In this chapter, we started building our custom blockchain in Rust. First, we planned out our entire project using visuals and laid out our structs and function plans. Then, we started writing the actual code that would help us create functions for creating blocks, transactions, and the genesis block, along with structs for proof-of-work consensus and transactions.

In the next chapter, we will take things further. Following our initially laid-out roadmap, we will build the rest of the features and functions. Just like we did in this chapter, we will go through all the lines of code so that you are clear about how the blockchain works.

4

Adding More Features to Our Custom Blockchain

In the previous chapter, we started our journey of building a blockchain using Rust. We created a visual plan of our structs and functions and also started writing some code. We looked at the basic building blocks of the blockchain project that we're building. We defined structs and some functions that help understand how many of the core components in the blockchain will be represented.

This chapter is the next step, where we extend our learning and go deeper into the implementation and understanding of the blockchain. We will look at the rest of the code, flesh out more details to the elements introduced in *Chapter 3*, *Building a Custom Blockchain*, and understand how these details work and contribute to the blockchain at large.

By the end of this chapter, we will have a blockchain program that compiles and runs.

We will cover the following topics:

- Connecting the blocks
- Starting the node server

Technical requirements

In this phase of advancing the custom blockchain introduced in the preceding chapters, essential features will be integrated into the existing implementation, significantly expanding its capabilities.

To ensure a seamless learning experience, all the code related to this chapter is available in the dedicated GitHub repository for our book at `https://github.com/PacktPublishing/Rust-for-Blockchain-Application-Development`.

You can clone the code with this command:

```
git clone https://github.com/PacktPublishing/Rust-for-Blockchain-
Application-Development.git
```

You'll find detailed code snippets, projects, and resources referenced from this repository.

For an immersive comprehension of the concepts discussed, it's highly recommended to clone this repository. By doing so, readers will seamlessly follow and interact with the provided code examples, enabling a firsthand observation of the blockchain's evolution.

Let's delve straight into the exploration and implementation.

Connecting the blocks

This section will start bringing important independent concepts together to build a functional blockchain.

Let us first start by looking at some essential libraries that will be used throughout this project.

Libraries powering blockchain operations

This section will cover crucial libraries essential for the `Blockchain` struct's operations. These libraries cover diverse aspects of the blockchain's core functionalities, enhancing its efficiency and overall performance.

Data storage with Sled

The `Blockchain` struct heavily relies on the Sled library for robust data storage management. By utilizing a key-value store approach, Sled is fully implemented in Rust and offers ACID transactions, which ensures data consistency even in the presence of errors such as crashes, and that's the right use case for our blockchain since data consistency is paramount. An interesting point to note is that Sled is an embedded database, meaning that it is designed to be directly integrated into applications without requiring a separate server process. It's also optimized for performance, employs zero-copy architecture, and provides various configuration options to customize its behavior according to the requirements of the application.

Sled becomes integral in persisting crucial blockchain components, including blocks and their associated transactions. This utilization of Sled ensures a streamlined and efficient method for storing and retrieving essential data within the blockchain.

Transaction and block handling

The integration of libraries such as `crate::transaction::TXOutput`, `crate::{Block, Transaction}`, and `sled::transaction::TransactionResult` facilitates seamless manipulation and handling of transactions and blocks within the `Blockchain` struct. These libraries contribute to the efficient organization and interaction of transactional data within the blockchain.

Encoding and decoding support

Utilizing `data_encoding::HEXLOWER` enables efficient encoding and decoding of data – in our case, in hexadecimal format. It makes this possible by providing utilities for encoding and decoding data using various binary-to-text encoding schemes, allowing developers to work with formats such as Base64, Base32, hexadecimal, and even custom encoding schemes, and this is a critical aspect for cryptographic operations and data representation within the blockchain.

It helps us by encoding binary data into a text-based format that is human-readable and easy to transmit over text-based protocols such as HTTP, JSON, or XML.

Data encoding is designed to be flexible and extensible and provides robust error-handling mechanisms to deal with invalid input data. One of the primary reasons that we have used this library is because of its performance and also the clear documentation available, along with clear examples and explanations of how to use its various features.

Key-value store implementation

The `sled::{Db, Tree}` libraries offer a scalable and efficient key-value storage system. Their integration provides the `Blockchain` struct with robust data storage capabilities, ensuring reliability and performance in managing blockchain components. We've already discussed Sled and how it helps us with consistent data storage – something that's required by our blockchain.

Data structure management

`std::collections::HashMap` plays a key role in organizing and managing data structures within the blockchain implementation. Its functionalities contribute to efficient data handling and retrieval within the blockchain. A **hashmap** essentially stores elements as key-value pair, where each key must be unique within the map, and this provides very efficient lookup operations, typically with an average time complexity of **O(1)** for inserting, updating, and retrieving elements. Hashmaps implement automatic resizing as needed to maintain efficient performance.

Filesystem interaction

The inclusion of `std::env::current_dir` enables seamless access to the current directory, facilitating file-handling operations crucial for blockchain data management and interaction. The `current_dir` function returns a result where the success variant contains a `PathBuf` pointer, which represents the working directory. The function is particularly useful when dealing with file I/O operations or when the program needs to know its current location within the filesystem. `PathBuf` here is a **smart pointer** to a path in the filesystem.

Concurrent access management

Utilizing `std::sync::{Arc, RwLock}` provides synchronization primitives, ensuring safe concurrent access to shared data structures. This is pivotal for maintaining data consistency and integrity in a multithreaded environment within the blockchain implementation. `Arc` here stands for **Atomic Reference Counting**; it is a thread-safe reference-counting pointer, allowing multiple threads to share ownership of a value. The purpose of `Arc` is to allow multiple threads to have read-only access to a value without the need for explicit locking – something that's very important in our blockchain application, where we'd like to support multiple concurrent reads without locking. `Arc` also maintains a reference count internally, which is incremented or decremented atomically as references to the value are acquired and released, and hence is used primarily for scenarios where you need to share immutable data across multiple threads – such as in our case.

`RwLock` stands for **reader-writer lock**; it provides a synchronization mechanism that allows multiple readers or a single writer to access a resource at the same time.

The lock allows concurrent read access, meaning multiple threads can acquire a read lock and access data simultaneously as long as no threads hold a write lock – this significantly helps with data consistency, and this is the same reason why we're using Sled as well.

Now, we'll move on to the blockchain functions that will help us perform operations on the blockchain.

Blockchain functions

Blockchain functions as a decentralized, secure ledger technology, facilitating transparent transactions. It encompasses consensus mechanisms, smart contracts, and tokenization. These functions enable diverse applications, from cryptocurrency transactions to supply chain traceability, offering a broad spectrum of decentralized solutions.

In the `blockchain.rs` file, readers will find a collection of essential functions designed to facilitate various operations within the blockchain. These functions are thoroughly explained here to ensure a comprehensive understanding. For those following along, the detailed code can be accessed within the GitHub repository.

create_blockchain function

The `blockchain::Blockchain::create_blockchain` function initializes a blockchain, utilizing the Sled key-value store. It checks for an existing blockchain and generates a genesis block if none is found. The resulting blockchain object includes a tip hash and database reference:

```
pub fn create_blockchain(genesis_address: &str) -> Blockchain {
    let db = sled::open(current_dir().unwrap().join("data")).unwrap();
    let blocks_tree = db.open_tree(BLOCKS_TREE).unwrap();
    let data = blocks_tree.get(TIP_BLOCK_HASH_KEY).unwrap();
    let tip_hash;
```

```
    if data.is_none() {
      let coinbase_tx = Transaction::new_coinbase_tx(genesis_address);
      let block = Block::generate_genesis_block(&coinbase_tx);
      Self::update_blocks_tree(&blocks_tree, &block);
      tip_hash = String::from(block.get_hash());
    } else {
      tip_hash = String::from_utf8(data.unwrap().to_vec()).unwrap();
    }
    Blockchain {
       tip_hash: Arc::new(RwLock::new(tip_hash)),
       db,
    }
  }
}
```

Overall, this function creates a new `Blockchain` instance by initializing a new database connection and creating a **genesis block** if there is no block in the blockchain yet. If there is already a block, it sets the `tip_hash` variable to the hash of the latest block.

The `genesis_address` parameter denotes the address designated to receive the initial reward for mining the first block on the blockchain. Let us now look at the algorithm of the function:

1. A new instance of the Sled embedded database library is created.

2. A database connection to a data directory is opened using the `current_dir()` and `join()` methods.

3. The `blocks_tree` key-value store in the database is opened using the `open_tree()` method.

4. An attempt is made to retrieve the value associated with the `TIP_BLOCK_HASH_KEY` key from the `blocks_tree` store using the `get()` method.

5. If the value is None, the process initiates the creation of a new genesis block by first establishing a **coinbase transaction** utilizing the specified `genesis_address` parameter. Subsequently, this transaction facilitates the generation of a fresh genesis block via `Block::generate_genesis_block()`. Following this creation, the `update_blocks_tree()` method is invoked to seamlessly insert this newly formed block into the `blocks_tree` store. Upon successful insertion, the `tip_hash` variable is assigned the hash value pertaining to this newly minted genesis block.

6. If the value is not None, the `tip_hash` variable is set to the value of the stored hash.

7. A new `Blockchain` instance is created with `tip_hash` wrapped in an `Arc<RwLock>` to allow for thread-safe access and the db instance.

The method does not explicitly return a value. Instead, actions such as creating a genesis block, updating the blockchain, and setting up a `Blockchain` instance for further use are performed. The return may be implicit, reflecting the initialized blockchain structure available for subsequent operations.

Now that we have learned how to create a `Blockchain` instance, let's discuss how to create a new blockchain.

new_blockchain function

The other function we need in the `blockchain.rs` file is the `blockchain::Blockchain::new_blockchain` function; let's take a look at what it does. The `new_blockchain` function initializes a blockchain, leveraging the Sled key-value store. It retrieves the latest block hash from storage, assuming a pre-existing blockchain. The resulting blockchain object includes a tip hash and a reference to the database:

```
pub fn new_blockchain() -> Blockchain {
    let db = sled::open(current_dir().unwrap().join("data")).unwrap();
    let blocks_tree = db.open_tree(BLOCKS_TREE).unwrap();
    let tip_bytes = blocks_tree
        .get(TIP_BLOCK_HASH_KEY)
        .unwrap()
        .expect("No existing blockchain found. Create one first.");
    let tip_hash = String::from_utf8(tip_bytes.to_vec()).unwrap();
    Blockchain {
        tip_hash: Arc::new(RwLock::new(tip_hash)),
        db,
    }
}
```

The function initiates a new instance of the Sled embedded database library. It opens a database connection to a data directory using the `current_dir()` and `join()` methods. Then, it does the following:

1. It opens the `blocks_tree` key-value store in the database using the `open_tree()` method.

2. It tries to get the value associated with the `TIP_BLOCK_HASH_KEY` key from the `blocks_tree` store using the `get()` method.

3. If the value is None, then it means that there is no block in the blockchain yet, and the function returns an error message using the `expect()` method.

4. If the value is not equal to None, it implies an existing block in the blockchain. In this case, the function sets the `tip_hash` variable to the value of the stored hash.

5. This conversion is achieved by using the `String::from_utf8()` method, which transforms a vector of bytes into a string.

6. It creates a new `Blockchain` instance with `tip_hash` wrapped in an `Arc<RwLock>` to allow thread-safe write operations and the db instance. The `Blockchain` instance is then returned.

Overall, this function creates a new `Blockchain` instance by initializing a new database connection and setting the `tip_hash` variable to the hash of the latest block in the blockchain. If there is no block in the blockchain yet, the function returns an error message.

Let us now move on and learn how we add some struct methods that will be used to access and modify the database instance and the tip hash of the blockchain in a thread-safe manner.

Adding struct methods

Readers will find the following struct methods within the `blockchain.rs` file in the GitHub repository. These functions serve critical roles within the struct, enabling access to the database, retrieving the tip hash, and setting a new tip hash for the blockchain structure. Let's talk about them briefly:

- `get_db(&self) -> &Db`: This method returns a reference to the `sled::Db` instance associated with the `Blockchain` instance. It does this by taking a reference to `self` and returning a reference to the db field:

  ```
  pub fn get_db(&self) -> &Db {
      &self.db
  }
  ```

- `get_tip_hash(&self) -> String`: This method returns the current tip hash of the blockchain as a string. It does this by obtaining a read lock on the `tip_hash` field using the `read()` method of the RwLock type, cloning the string value using the `clone()` method, and then returning the cloned string:

  ```
  pub fn get_tip_hash(&self) -> String {
      self.tip_hash.read().unwrap().clone()
  }
  ```

- `set_tip_hash(&self, new_tip_hash: &str)`: This method sets the tip hash of the blockchain to a new value. It does this by obtaining a write lock on the `tip_hash` field using the `write()` method of the RwLock type, updating the value of the string by assigning it the new value passed as a parameter, and then releasing the write lock. The value passed as a parameter is first converted to a string using the `String::from()` method:

  ```
  pub fn set_tip_hash(&self, new_tip_hash: &str) {
      let mut tip_hash = self.tip_hash.write().unwrap();
      *tip_hash = String::from(new_tip_hash)
  }
  ```

Overall, these methods provide a way to access and modify the database instance and the tip hash of the blockchain in a thread-safe manner. The `get_db()` method allows external code to access the underlying database, while the `get_tip_hash()` and `set_tip_hash()` methods provide ways to access and modify the current tip hash of the blockchain.

Following this, readers will encounter an `iterator` function. This function plays a crucial role in initializing a `BlockchainIterator` struct. It utilizes the tip hash and database to create an iterator, facilitating traversal through the blockchain structure:

```
pub fn iterator(&self) -> BlockchainIterator {
    BlockchainIterator::new(self.get_tip_hash(), self.db.clone())
}
```

Here's how it works:

1. The method calls `self.get_tip_hash()` to retrieve the current tip hash of the blockchain.

2. It then calls the `clone()` method on the `self.db` instance to create a new instance of the `sled::Db` database, because the `BlockchainIterator` struct requires ownership of the database instance.

3. Finally, the method creates a new instance of the `BlockchainIterator` struct by passing the current tip hash and the cloned database instance as parameters to its `new()` method. The resulting iterator is then returned. Overall, this method returns a new iterator over the blocks in the blockchain by creating a new `BlockchainIterator` instance and passing it the current tip hash and a cloned instance of the database. This allows external code to iterate over the blocks in the blockchain in a thread-safe and efficient manner.

Let's quickly talk about the blockchain iterator since we used it in this method.

BlockchainIterator function

We don't yet have the struct or implementation for the `BlockchainIterator` struct. Let's work on both of these now:

*   The blockchain iterator simply helps us traverse the blockchain, facilitating sequential navigation through its blocks and its data. Its functionality and implementation details can be observed in the file when we delve into the implementation of the `BlockchainIterator` struct:

    ```
    pub struct BlockchainIterator {
        db: Db,
        current_hash: String,
    }

    impl BlockchainIterator {
        fn new(tip_hash: String, db: Db) -> BlockchainIterator {
            ...
        }

        pub fn next(&mut self) -> Option<Block> {
            ...
    ```

```
        }
    }
```

- The `BlockchainIterator::new` method takes `tip_hash`, representing the latest block's hash, and Db, a reference to the blockchain database (Db struct). The method facilitates blockchain navigation using the `next()` method, which retrieves blocks from the database.

- Within `next()`, the algorithm involves the following:

    I.   Opening `BLOCKS_TREE` within the database.

    II. Fetching block data based on `current_hash`.

    III. Deserializing the retrieved data into a `Block` object.

    IV. Updating `current_hash` to the previous block's hash.

    V.   Returning the deserialized block if data is found for the current hash; otherwise, returning `None`.

- The `next()` method returns an `Option<Block>` instance which results in either of the two following scenarios:

    I.   If data is successfully retrieved and a block is constructed, `Some(block)` with the deserialized block is returned.

    II. If no data is found for the current hash, `None` is returned, indicating the end of the blockchain iteration.

Overall, this code provides an iterator-like interface for iterating over the blocks in a blockchain stored in a database represented by a Db struct.

Let us now move on to the update of block trees and mining of blocks.

Mining blocks and updating block trees

New blocks need to be mined, and only then they will be added to a blockchain; after blocks are added, the block trees need to be updated as well. So, in this part, we will tackle the important job of mining the blocks and updating the block trees.

The next function in the file illustrates how the `mine_block` function creates a new block and incorporates it into the represented blockchain within the current instance of the `Blockchain` struct:

```
pub fn mine_block(&self, transactions: &[Transaction]) -> Block {
    for trasaction in transactions {
        if trasaction.verify(self) == false {
            panic!("ERROR: Invalid transaction")
        }
    }
}
```

```
    let best_height = self.get_best_height();
    let block = Block::new_block(self.get_tip_hash(), transactions,
best_height + 1);
    let block_hash = block.get_hash();
    let blocks_tree = self.db.open_tree(BLOCKS_TREE).unwrap();
    Self::update_blocks_tree(&blocks_tree, &block);
    self.set_tip_hash(block_hash);
    block
  }
```

This function accepts a slice of `Transaction` structs as input. It then iterates through each transaction, verifying them individually using the `verify()` method of the `Transaction` struct. If any transaction is found to be invalid during this process, the function panics, providing an error message.

Here's how it works:

1. The function begins by retrieving the height of the highest block in the blockchain via `self.get_best_height()`.

2. Using the `Block::new_block()` method, a new block is created by passing the following:

    *   Transactions awaiting inclusion in the block.

    *   The hash of the tip block.

    *   The height of the new block, which is one higher than the best height.

    *   The *best height* signifies the highest block index in the blockchain, indicating the current position of the latest block in the chain. It's pivotal in establishing the new block's position within the existing chain, ensuring its integrity and continuity.

3. Subsequently, the hash of the newly created block is computed using the `get_hash()` method of the `Block` struct.

4. The function proceeds to access `BLOCKS_TREE` within the database: `self.db.open_tree(BLOCKS_TREE).unwrap()`.

5. It updates the tree with the information of the newly created block. However, the method responsible for this update, `update_blocks_tree()`, is not shown in the provided code snippet. This method is a static one and must be explicitly invoked within the `mine_block()` function to update `BLOCKS_TREE`.

6. Following the update, the function sets the tip hash of the blockchain to the hash of the new block using `self.set_tip_hash(block_hash)`. Finally, the function returns the newly created block.

Moving on to the update process at the start of the file, readers will find a crucial function named `update_blocks_tree`. This function serves the purpose of updating BLOCKS_TREE within the database:

```
fn update_blocks_tree(blocks_tree: &Tree, block: &Block) {
    let block_hash = block.get_hash();
    et _: TransactionResult<(), ()> = blocks_tree.transaction(|tx_db|
{
    let _ = tx_db.insert(block_hash, block.clone());
    let _ = tx_db.insert(TIP_BLOCK_HASH_KEY, block_hash);
    Ok(())
    });
}
```

This `update_blocks_tree` function updates the `blocks_tree` database tree with the new `Block` object passed as an argument:

1. The function initiates by computing the hash of the `Block` object using the `get_hash()` method within the `Block` struct. It utilizes a transactional approach on `blocks_tree`, employing the `transaction()` method from the `sled::Tree` struct.

2. The `TransactionResult` type encapsulates the outcome of the transaction. If the transaction is successful, it yields `Ok(())`. In case of a failure, it returns `Err(())`.

3. Within the transaction, the function inserts two distinct key-value pairs into the tree:

    I.   First, it inserts the `Block` object with the `block_hash` key using `tx_db.insert(block_hash, block.clone())`. This addition involves a new key-value pair, where the key represents the hash of the block and the value is a serialized version of the block.

    II. Secondly, it inserts the `TIP_BLOCK_HASH_KEY` key-value pair into the tree via `tx_db.insert(TIP_BLOCK_HASH_KEY, block_hash)`. This action associates the value of `TIP_BLOCK_HASH_KEY` with the hash of the newly added block.

4. Upon successful completion, the function returns `Ok(())`, signifying a successful transaction. Conversely, if the transaction fails, it returns `Err(())`.

Overall, this function updates the `blocks_tree` database tree by adding a new block to it and updating the tip block hash.

Now, we will add some blocks to our blockchain and look at some more helper functions.

Adding blocks and some helper functions

First, readers will find the `add_block` function in the file. This function adds the block to the blockchain after it's mined. Following this, there are some helper functions that the `add_block` function calls upon to help it with particular tasks.

The `add_block` function accepts a reference to a `Block` object as its input parameter. It operates within the context of the `Blockchain` struct, utilizing the database to manage the blockchain's block storage:

```
pub fn add_block(&self, block: &Block) {
    let block_tree = self.db.open_tree(BLOCKS_TREE).unwrap();
    if let Some(_) = block_tree.get(block.get_hash()).unwrap() {
      return;
    }
    let _: TransactionResult<(), ()> = block_tree.transaction(|tx_db|
{
    let _ = tx_db.insert(block.get_hash(), block.serialize()).
unwrap();

    let tip_block_bytes = tx_db
        .get(self.get_tip_hash())
        .unwrap()
        .expect("The tip hash is not valid");
    let tip_block = Block::deserialize(tip_block_bytes.as_ref());
      if block.get_height() > tip_block.get_height() {
    let _ = tx_db.insert(TIP_BLOCK_HASH_KEY, block.get_hash()).
unwrap();
        self.set_tip_hash(block.get_hash());
    }

        Ok(())
    });
}
```

The function starts by opening BLOCKS_TREE within the database to access block-related information.

It checks if the block already exists within the tree by retrieving its hash. If the block's hash is found in the tree, indicating an existing block, the function exits. Within a transactional operation, it does the following:

1. It inserts the block's serialized form into the tree, associating it with its hash.

2. It retrieves the bytes of the block referenced by the tip hash.

3. It deserializes this block to compare heights with the incoming block.

4. If the incoming block has a greater height than the tip block, it updates the tip block's hash in the tree and sets the blockchain's tip hash accordingly.

5. Upon successful execution, the function concludes with a successful result, (Ok(())).

Overall, the add_block function serves to incorporate a new block into the blockchain stored within the database. It verifies the absence of a block in the tree before inserting one and manages the blockchain's tip block, ensuring the insertion of blocks in sequential order and maintaining the blockchain's integrity and chronological structure.

Next, readers will encounter three helper functions that are being used not only by the add_block function but have also been called by the mine_block and the update_block_tree functions.

Within this context, three distinct functions emerge: get_best_height, get_block, and get_block_hashes. These functions, serving helpers as mentioned earlier, play important roles in our program. Their individual contributions and functionalities will now be explored:

1. Starting with the get_best_height function, readers will encounter this function first within the file. This function returns the height of the block with the highest height in the blockchain. It does this by first getting the serialized bytes of the block corresponding to the current tip hash, deserializing it into a Block struct, and returning its height.

2. Next, the get_block function takes a block hash as input, retrieves the corresponding serialized block bytes from the blockchain database, deserializes them into a Block struct, and returns the block. If the block with the given hash is not found in the database, it returns None.

3. The final function that readers will encounter is the get_block_hashes function; this function returns a list of all block hashes in the blockchain. It does this by iterating over all blocks in the blockchain using the iterator function, getting the hash of each block, converting it to a byte vector, and adding it to a list. The function returns this list of block hashes.

Let us now understand **unspent transaction outputs' (UTXOs')** significance and transaction retrieval's role within the blockchain structure for comprehensive understanding.

Finding UTXOs and transactions

To finish off the blockchain.rs file, readers will find the find_utxo and find_transaction functions; let's go through them one by one.

Let's look at the fint_utxo function in the file and then break it down to understand what it does:

```
pub fn find_utxo(&self) -> HashMap<String, Vec<TXOutput>> {
    let mut utxo: HashMap<String, Vec<TXOutput>> = HashMap::new();
    let mut spent_txos: HashMap<String, Vec<usize>> = HashMap::new();

    let mut iterator = self.iterator();
    loop {
```

```
    let option = iterator.next();
      if option.is_none() {
      break;
    }
    let block = option.unwrap();
      'outer: for tx in block.get_transactions() {
    let txid_hex = HEXLOWER.encode(tx.get_id());
        for (idx, out) in tx.get_vout().iter().enumerate() {


          }
        }
      }
    utxo
  }
```

The find_utxo function, the concluding component of our file, operates with no input parameters. It navigates through the blockchain, identifying UTXOs by inspecting each transaction within each block. Let's break down its functionality:

1. The find_utxo function initializes two HashMap: utxo instances to store UTXOs and spent_txos to track **spent TXOs (STXOs)**.

2. It starts an iteration through blocks using an iterator from the self object.

3. Within the loop, it retrieves the next block until there are no more blocks left. For each block, it enters a nested loop, iterating through each transaction (tx) within the block.

4. It encodes the transaction ID (txid_hex) into a hexadecimal string using a HEXLOWER encoding method.

5. Inside the nested loop, it further iterates through each output (out) of the transaction and checks if the output is spent by referencing spent_txos based on the transaction ID and output index.

6. If the output is found in spent_txos, it skips to the next transaction using a labeled loop (outer).

7. If the output is not spent, it checks if the utxo hashmap already contains the transaction ID. If so, it appends the output to the corresponding vector. Otherwise, it creates a new entry in the utxo hashmap. Additionally, if the transaction is a coinbase transaction (newly generated coins), it skips further processing for that transaction.

8. Moving on, it loops through each input (txin) of the transaction and extracts the referenced transaction ID (txid_hex) from it.

9. It checks if spent_txos already contains the referenced transaction ID. If found, it appends the input's output index to the corresponding vector. Otherwise, it creates a new entry in spent_txos. The function continues this process until it exhausts all blocks and transactions, finally returning the populated utxo hashmap containing UTXOs.

The find_utxo function is essential for analyzing and managing UTXOs within the blockchain, ensuring accurate tracking of unspent outputs for transaction validity and integrity.

Now, let's go ahead and look at the find_transaction function and understand what it does.

The find_transaction function searches the blockchain for a specific transaction by its transaction ID. Utilizing an iterative approach, it navigates through the entire blockchain, examining each block and its contained transactions to identify and retrieve the desired transaction if it exists. If a transaction with the provided ID is found, it is cloned and returned, providing a comprehensive means of transaction identification within the blockchain data structure:

```
pub fn find_transaction(&self, txid: &[u8]) -> Option<Transaction> {
    let mut iterator = self.iterator();
      loop {
    let option = iterator.next();
      if option.is_none() {
        break;
    }
    let block = option.unwrap();
        for transaction in block.get_transactions() {
            if txid.eq(transaction.get_id()) {
            return Some(transaction.clone());
                }
            }
        }
        None
    }
```

Here's how it works:

1. The find_transaction function searches through all the transactions in the blockchain for a transaction with a given transaction ID (txid) and returns that transaction if found, or None otherwise.

2. The function starts by creating an iterator over the blockchain, which allows it to loop through all blocks in the blockchain. It then enters a loop where it calls the next() method on the iterator to get the next block in the chain, and checks whether the option is None. If it is, that means there are no more blocks left to iterate over, and the loop breaks.

3. For each block, the function loops through all transactions in the block using the get_transactions() method. For each transaction, it checks whether the transaction ID (txid) matches the ID of the current transaction being iterated over using the get_id() method. If there is a match, the function returns a clone of the transaction using the clone() method.

4. If the loop completes without finding a matching transaction, the function returns None.

This function serves as a tool to retrieve transactions based on their unique IDs within the blockchain structure. With this function, we have completed our exploration of the `blockchain.rs` file. In the next sections, we will tackle each of the concepts we used here separately – transactions, UTXOs, nodes, servers, and memory pool.

Moving to the next section, we will see how to start up a node server.

Starting the node server

As we learned in the first chapter, a copy of the blockchain is maintained on the nodes that participate in the network, and this is what makes blockchains decentralized since a copy of the blockchain exists on multiple nodes, each of which acts as a server. Even if one node goes down or a copy is deleted, other nodes are still there to uphold the blockchain network.

So far, we have built the blocks, mined them, added them, and connected them to the blockchain by updating the block tree. Now, it's time to see how blockchain copies are maintained on the nodes, and before that, we will need to start a server.

In this section, we will go through some code that will help us start servers and help operate nodes for our blockchain network.

The server

In the project's GitHub repository, readers will discover a file named `server.rs`, dedicated to hosting the necessary code for initializing a server.

The crates that you see in the first line are the various crates we have either previously discussed or will be further elaborated on shortly – for example, block, nodes, memory pool, transaction, global config, and so on.

The readers will then encounter some standard libraries used in Rust development, such as `time`, `io`, `net`, `error`, and `thread`. These libraries streamline common tasks, offering pre-built functions that eliminate the need to write certain functionalities from scratch. They provide essential tools for handling errors, managing time, dealing with I/O operations, networking, and concurrent threading. Documentation for these libraries is readily available in the standard Rust library documentation.

We also see libraries such as `serde` that help us to work with JSON data by serializing and deserializing the data from structs to JSON and vice versa.

Let's understand this better:

- `Block`, `Blockchain`, `Transaction`, and `UTXOSet` are custom types defined in the same crate as this module.

- `BlockInTransit`, `MemoryPool`, and `Nodes` are custom types defined in other modules within the same crate.

- `data_encoding::HEXLOWER` is a crate that provides utilities for encoding and decoding hexadecimal data.

- `log::{error, info}` is a crate for logging error and info messages.

- `once_cell::sync::Lazy` is a crate that provides a lazy evaluation macro.

- `serde::{Deserialize, Serialize}` is a crate for serializing and deserializing Rust data structures.

- `serde_json::Deserializer` is a deserializer for JSON data structures.

- `std::error::Error` is a trait for errors in Rust.

- `std::io::{BufReader, Write}` provides I/O support for reading and writing data.

- `std::net::{Shutdown, SocketAddr, TcpListener, TcpStream}` provides support for network programming in Rust.

- `std::thread` provides support for concurrent programming in Rust.

- `std::time::Duration` provides support for representing time durations in Rust.

- `NODE_VERSION` is an unsigned integer that represents the version of the cryptocurrency node.

- `CENTERAL_NODE` is a string that represents the IP address and port number of the central node that the new node will connect to.

- We also define a static lazy evaluated variable named `GLOBAL_NODES` that is of type `Nodes`. This variable is used to store the network nodes that this node is aware of. The `Lazy` type allows for the creation of the variable to be deferred until it is needed. The closure passed to `Lazy::new()` initializes `GLOBAL_NODES` by creating a new `Nodes` instance and adding the `CENTERAL_NODE` string to it.

- We then define a static lazy evaluated variable named `GLOBAL_MEMORY_POOL` that is of type `MemoryPool`. This variable is used to store pending transactions that have not yet been included in a block. The closure passed to `Lazy::new()` initializes `GLOBAL_MEMORY_POOL` by creating a new `MemoryPool` instance.

Now, we will explore the code for the `Server` struct and the implementation of some other methods.

Server struct and implemented methods

Here is the `Server` struct, which defines essential functionalities to handle incoming client connections, initiate communication with a central node, and concurrently manage requests from multiple clients through separate threads. Through this implementation, the server efficiently interacts with the blockchain, utilizing the `serve` function to handle diverse client requests while maintaining synchronization and robustness within the network.

The `Server` struct contains a `blockchain` field of type `Blockchain`:

```
pub struct Server {
    blockchain: Blockchain,
}

impl Server {
    pub fn new(blockchain: Blockchain) -> Server {
        Server { blockchain }
    }

    pub fn run(&self, addr: &str) {
        let listener = TcpListener::bind(addr).unwrap();

        if addr.eq(CENTERAL_NODE) == false {
            let best_height = self.blockchain.get_best_height();
            send_version(CENTERAL_NODE, best_height);
        }
        for stream in listener.incoming() {
            let blockchain = self.blockchain.clone();
            thread::spawn(|| match stream {
                Ok(stream) => {

                }
                Err(e) => {

                }
            });
        }
    }
}
```

The implementation of `Server` includes a new associated function that initializes a new `Server` instance with a provided `Blockchain` type.

The `run()` method takes a mutable reference to self (`&self`) and an address (`&str`) to start the server. Here's how it works:

1. It creates a TCP listener bound to the provided address.
2. If the provided address is not equal to a predefined constant, `CENTERAL_NODE`, it retrieves the best height of the blockchain and sends a version message to the central node.
3. Regardless of the conditional check, it logs the initiation of the node server at the given address.
4. The function enters a loop to handle incoming TCP streams from the listener.

5. For each incoming stream, it captures a clone of blockchain for use within the spawned thread.

6. It creates a new thread for each incoming connection using thread::spawn.

7. Within the spawned thread, it attempts to match the incoming stream:

8. If the stream is successfully received (Ok(stream)), it tries to serve the client by calling the serve function with the captured Blockchain instance and the stream.

9. If an error occurs during the Err(e) stream handling, it logs a connection failure.

10. This loop continues to listen for and handle incoming connections in separate threads, ensuring the server can handle multiple concurrent connections.

The Server struct listens for incoming connections on the specified address and handles each connection in a new thread by calling the serve function. The serve function handles requests from clients using the provided Blockchain instance.

Moving on, let us now discuss some enums.

Enums

In the server code, meaning the server.rs file in the repository, we can see a serve function that will actually serve the entire blockchain; this function needs enums and helper functions. Let's go ahead and look at the required enums.

In the code within the file, readers will come across two essential enums: OpType and Package. These enums play an important role in organizing and transmitting serialized data across the blockchain network.

First, let us look at the OpType enum, which encapsulates the fundamental types of operations pertinent to blockchain functionality. The Tx variant signifies operations related to transactions, while the Block variant represents activities linked to blocks within the blockchain.

Package is a more complex enum with six possible values, each of which contains some associated data. The possible values are the following:

- Block: Contains an addr_from field (a String type) and a block field (a Vec<u8> type)

- GetBlocks: Contains an addr_from field (a String type)

- GetData: Contains an addr_from field (a String type), an op_type field (an OpType type), and an id field (a Vec<u8> type)

- Inv: Contains an addr_from field (a String type), an op_type field (an OpType type), and an items field (a Vec<Vec<u8>>> type)

- Tx: Contains an addr_from field (a String type) and a transaction field (a Vec<u8> type)

- Version: Contains an addr_from field (a String type), a version field (a usize type), and a best_height field (a usize type)

The #[derive(Debug, Serialize, Deserialize)] annotations above each enum indicate that Debug, Serialize, and Deserialize traits should be automatically implemented for these types. This allows the enums to be printed with println!("{:?}", some_enum) and to be serialized/deserialized to and from bytes using a Rust serialization library such as serde.

In preparing for the serve function, let's understand some essential helper functions necessary to support its functionalities in the next section.

Helper functions

Before we look at the serve function, we need to understand some helper functions that'll be used in the serve function, so in this section, let's go ahead and do just that.

In total, we need seven helper functions. Let us look at them one at a time.

send_get_data function

This function transmits a request for specific data to a designated network address:

```
fn send_get_data(addr: &str, op_type: OpType, id: &[u8]) {
    let socket_addr = addr.parse().unwrap();
    let node_addr = GLOBAL_CONFIG.get_node_addr().parse().unwrap();
    send_data(
        socket_addr,
        Package::GetData {
            addr_from: node_addr,
            op_type,
            id: id.to_vec(),
        },
    );
}
```

Here's how it works:

1. First, the function takes in an address (addr: &str), an operation type (op_type: OpType), and an ID in the form of a byte slice (id: &[u8]).

2. It parses the addr input into a SocketAddr structure using the parse() method and unwraps the result.

3. It retrieves the node address from a global configuration (GLOBAL_CONFIG) and parses it into a SocketAddr structure, also unwrapping the result.

4. It then invokes a `send_data` function, passing in the following:

    *   The parsed `socket_addr` function as the destination address for sending data.

    *   A `Package::GetData` enum variant containing the following:

        *   `addr_from`: The parsed `node_addr` function, representing the address of the node sending the data

        *   `op_type`: The specified operation type

        *   `id.to_vec()`: The ID converted into a `Vec<u8>` type, which is included in the package for data retrieval

This function abstracts the process of sending a specific type of data (`GetData`) to a specified address using a standardized package format, including source address, operation type, and an associated ID. This function will initiate a data retrieval request to the specified address in the blockchain network.

send_inv function

This function notifies us about specific data items to a provided network address:

```
fn send_inv(addr: &str, op_type: OpType, blocks: &[Vec<u8>]) {
    let socket_addr = addr.parse().unwrap();
    let node_addr = GLOBAL_CONFIG.get_node_addr().parse().unwrap();
    send_data(
        socket_addr,
        Package::Inv {
            addr_from: node_addr,
            op_type,
            items: blocks.to_vec(),
        },
    );
}
```

Here's how it works:

1. The function takes an address (`addr: &str`), an operation type (`op_type: OpType`), and a slice of byte vectors representing blocks (`blocks: &[Vec<u8>]`).

2. It parses the `addr` input into a `SocketAddr` structure using the `parse()` method and unwraps the result.

3. It retrieves the node address from a global configuration (`GLOBAL_CONFIG`) and parses it into a `SocketAddr` structure, also unwrapping the result.

4. It then calls a `send_data` function, passing in the following:

   - The parsed `socket_addr` function as the destination address for sending data.

   - A `Package::Inv` enum variant containing the following:

     - `addr_from`: The parsed `node_addr` function, representing the address of the node sending the data

     - `op_type`: The specified operation type

     - `blocks.to_vec()`: The blocks are converted into a `Vec<Vec<u8>>` type, which is included in the package as items to be communicated

This function abstracts the process of sending inventory information (`Inv`) to a specified address using a standardized package format, including source address, operation type, and a collection of byte vector items, which in this case represent blocks. This function will help broadcast inventory notifications for specific data items to the indicated network address.

send_block function

This function transmits a block to a specified network address:

```
fn send_block(addr: &str, block: &Block) {
    let socket_addr = addr.parse().unwrap();
    let node_addr = GLOBAL_CONFIG.get_node_addr().parse().unwrap();
    send_data(
        socket_addr,
        Package::Block {
            addr_from: node_addr,
            block: block.serialize(),
        },
    );
}
```

Here's how it works:

1. The `send_block` function accepts an address (`addr: &str`) and a reference to a `Block` instance (`block: &Block`).

2. It parses the `addr` input into a `SocketAddr` structure using the `parse()` method and unwraps the result.

3. It retrieves the node address from a global configuration (`GLOBAL_CONFIG`) and parses it into a `SocketAddr` structure, also unwrapping the result.

4. The function then invokes a `send_data` function, passing in the following:

    - The parsed `socket_addr` function as the destination address for sending data.

    - A `Package::Block` enum variant containing the following:

      - `addr_from`: The parsed `node_addr` function, representing the address of the node sending the data

      - `block.serialize()`: The serialized form of the `Block` struct obtained by invoking the `serialize()` method on the provided block

This function abstracts the process of sending a block (`Block`) to a specified address using a standardized package format. The block is serialized before sending, likely to transmit it efficiently in byte form over the network.

send_tx function

This function dispatches a transaction to a specified network address:

```
pub fn send_tx(addr: &str, tx: &Transaction) {
    let socket_addr = addr.parse().unwrap();
    let node_addr = GLOBAL_CONFIG.get_node_addr().parse().unwrap();
    send_data(
        socket_addr,
        Package::Tx {
            addr_from: node_addr,
            transaction: tx.serialize(),
        },
    );
}
```

Here's how it works:

1. The `send_tx` function takes an address (`addr: &str`) and a reference to a `Transaction` instance (`tx: &Transaction`).

2. It parses the `addr` input into a `SocketAddr` structure using the `parse()` method and unwraps the result.

3. It retrieves the node address from a global configuration (`GLOBAL_CONFIG`) and parses it into a `SocketAddr` structure, also unwrapping the result.

4.   The function then invokes a `send_data` function, passing in the following:

- The parsed `socket_addr` function as the destination address for sending data.

- A `Package::Tx` enum variant containing the following:

  - `addr_from`: The parsed `node_addr` function, representing the address of the node sending the data

  - `tx.serialize()`: The serialized form of the `Transaction` struct obtained by invoking the `serialize()` method on the provided `tx` instance

This function abstracts the process of sending a transaction (`Transaction`) to a specified address using a standardized package format. The transaction is serialized before sending, presumably for efficient transmission over the network.

send_version function

This broadcasts version information to a specified network address:

```
fn send_version(addr: &str, height: usize) {
    let socket_addr = addr.parse().unwrap();
    let node_addr = GLOBAL_CONFIG.get_node_addr().parse().unwrap();
    send_data(
        socket_addr,
        Package::Version {
            addr_from: node_addr,
            version: NODE_VERSION,
            best_height: height,
        },
    );
}
```

Here's how it works:

1.   The `send_version` function takes an address (`addr: &str`) and a height (`height: usize`) as parameters.

2.   It parses the `addr` input into a `SocketAddr` structure using the `parse()` method and unwraps the result.

3.   It retrieves the node address from a global configuration (`GLOBAL_CONFIG`) and parses it into a `SocketAddr` structure, also unwrapping the result.

4.   The function then invokes a `send_data` function, passing in the following:

- The parsed `socket_addr` function as the destination address for sending data.

- A `Package::Version` enum variant containing the following:

    - `addr_from`: The parsed `node_addr` function, representing the address of the node sending the data

    - `version`: A `NODE_VERSION` constant representing the version of the node's protocol or software

    - `best_height`: The provided height, indicating the best height or the latest block height known to the sending node

This function abstracts the process of sending a version message (`Package::Version`) to a specified address using a standardized package format. The version message includes information about the node's version and the best-known block height.

send_get_blocks function

This transmits a request for block data to a specified network address:

```
fn send_get_blocks(addr: &str) {
    let socket_addr = addr.parse().unwrap();
    let node_addr = GLOBAL_CONFIG.get_node_addr().parse().unwrap();
    send_data(
        socket_addr,
        Package::GetBlocks {
            addr_from: node_addr,
        },
    );
}
```

Here's how it works:

1. The `send_get_blocks` function takes an address (`addr: &str`) as its parameter.

2. It parses the `addr` input into a `SocketAddr` structure using the `parse()` method and unwraps the result.

3. It retrieves the node address from a global configuration (`GLOBAL_CONFIG`) and parses it into a `SocketAddr` structure, also unwrapping the result.

4. The function then invokes a `send_data` function, passing in the following:

    - The parsed `socket_addr` function as the destination address for sending data.

    - A `Package::GetBlocks` enum variant containing the following:

        - `addr_from`: The parsed `node_addr` function, representing the address of the node sending the data.

This function abstracts the process of sending a request for blocks (`Package::GetBlocks`) to a specified address using a standardized package format. This request does not include any specific block IDs or other parameters; it simply requests blocks from the receiving node.

send_data Function

This function sends data packages to a specified socket address:

```
fn send_data(addr: SocketAddr, pkg: Package) {
    info!("send package: {:?}", &pkg);
    let stream = TcpStream::connect(addr);
    if stream.is_err() {
        error!("The {} is not valid", addr);

        GLOBAL_NODES.evict_node(addr.to_string().as_str());
        return;
    }
    let mut stream = stream.unwrap();
    let _ = stream.set_write_timeout(Option::from(Duration::from_
millis(TCP_WRITE_TIMEOUT)));
    let _ = serde_json::to_writer(&stream, &pkg);
    let _ = stream.flush();
}
```

Here's how it works:

1. The `send_data` function takes `SocketAddr` and `Package` as parameters.

2. It logs information about the package being sent using the `info!` macro.

3. It attempts to establish a TCP connection (`TcpStream::connect(addr)`).

4. If the connection attempt fails (`stream.is_err()`), it logs an error message, evicts the node associated with the invalid address from `GLOBAL_NODES`, and returns from the function.

5. If the connection is successful, it proceeds with the established stream.

6. It sets a write timeout for the stream using `stream.set_write_timeout()`.

7. It serializes the `Package` parameter (pkg) using `serde_json::to_writer()` and sends it over the stream.

8. The function then flushes the stream to ensure all data is written.

Essentially, this function abstracts the process of sending a package (`Package`) over a TCP connection to a specified socket address, handling errors and serialization of data.

All these functions abstract away the details of sending different types of packages over a network connection and provide a simple interface for the caller to send data.

Now, we'll outline the `serve` function, a pivotal part of the blockchain project's server functionality, providing a detailed explanation of its intricate code structure.

The serve function

The `serve` function is a function that is more than 100 lines of code, and this is why we will mention just the outline of this function here instead of explaining it in detail. The actual code for this function can be referred to from the official GitHub repository associated with this book. In the blockchain project, in the `server.rs` file, toward the middle of the file, you will find the `serve` function. This is what it looks like:

```
fn serve(blockchain: Blockchain, stream: TcpStream) -> Result<(),
Box<dyn Error>> {

let _ = stream.shutdown(Shutdown::Both);
    Ok(())
}
```

This function receives a TCP stream connection and a `Blockchain` instance. It deserializes incoming packages from the stream and processes them based on their type. Here is a brief summary of what the function does for each type of package:

- For `Package::Block`, it deserializes the block and adds it to the blockchain. If there are blocks in transit, it sends a `get_data` request for the next block. If there are no more blocks in transit, it reindexes the UTXO set of the blockchain.

- For `Package::GetBlocks`, it retrieves all block hashes from the blockchain and sends an `inv` message with a list of hashes to the requesting peer.

- For `Package::GetData`, it retrieves the requested block or transaction from the blockchain or the global memory pool and sends it back to the requesting peer.

- For `Package::Inv`, it adds the received blocks or transactions to the global blocks in transit or the memory pool and requests missing blocks or transactions via `get_data` if necessary.

- For `Package::Tx`, it deserializes the transaction and adds it to the global memory pool. If the node is a miner and the memory pool has reached a certain threshold, it creates a new block containing transactions from the memory pool, mines it, and broadcasts the new block to other nodes via `inv`.

- For `Package::Version`, it compares the heights of the local and the remote blockchain and sends a `get_blocks` or `version` message to the remote node if necessary. It also adds the remote node to the global list of known nodes if it is not already present.

The function returns a `Result` indicating success or failure and closes the stream connection after processing all incoming packages.

Moving on, let us now explore the Node struct, which we will use to serve the blockchain.

The Node struct

We already have the logic for the server, and the node is what's serving the blockchain. Let's understand some code that is used to implement the node. For this, readers are advised to refer to the node.rs file:

```
#[derive(Clone)]
pub struct Node {
    addr: String,
}
impl Node {
    fn new(addr: String) -> Node {
        Node { addr }
    }
    pub fn get_addr(&self) -> String {
        self.addr.clone()
    }
    pub fn parse_socket_addr(&self) -> SocketAddr {
        self.addr.parse().unwrap()
    }
}
```

The code defines a Node structure used to represent network nodes in the blockchain. The Node structure has a single field, addr, which stores the node's address. It includes methods for creating a new Node instance, retrieving the node's address, and parsing the address into a SocketAddr structure. This allows easy access to the node's address and facilitates conversion for network communication within the blockchain protocol.

The Node struct has a single addr field of type String to store the address. Let's look at this in more detail:

1. The implementation of Node includes a new associated function to create a new Node instance with a provided address.

2. The get_addr() method returns a clone of the stored address (addr) as a String type.

3. The parse_socket_addr() method attempts to parse the stored address (addr) into a SocketAddr structure.

4. If successful, it returns the parsed SocketAddr structure.

This struct provides a structured representation of a blockchain network node containing its address.

Now, let us understand some code that we will need to manage multiple nodes:

```
pub struct Nodes {
    inner: RwLock<Vec<Node>>,
}
impl Nodes {
    pub fn new() -> Nodes {
        Nodes {
            inner: RwLock::new(vec![]),
        }
    }
    pub fn add_node(&self, addr: String) {
        ...
    }
    pub fn evict_node(&self, addr: &str) {
        ...
    }
    pub fn first(&self) -> Option<Node> {
        ...
    }
    pub fn get_nodes(&self) -> Vec<Node> {
        ...
    }
    pub fn len(&self) -> usize {
        ...
    }
    pub fn node_is_known(&self, addr: &str) -> bool {
        ...
    }
}
```

The code defines a Nodes struct that represents a collection of Node objects. The collection is implemented using a RwLock<Vec<Node>>, which allows multiple readers or a single writer to access the collection simultaneously.

The Nodes struct has several methods:

- new() creates a new empty Nodes object

- add_node(addr: String) adds a new Node object to the collection with the given address, addr, but only if the address is not already in the collection

- evict_node(addr: &str) removes the Node object with the given address, addr, from the collection if it exists

- `first()` returns the first Node object in the collection, or None if the collection is empty

- `get_nodes()` returns a copy of the entire collection as a Vec<Node> collection

- `len()` returns the number of Node objects in the collection

- `node_is_known(addr: &str)` returns true if a Node object with the given address, addr, is in the collection and false otherwise

Each method of the Nodes struct acquires a read or write lock on the inner field, which is the underlying Vec<Node> collection. If a method acquires a write lock, it can modify the collection, while methods that acquire read locks can only read from the collection. The unwrap() method is used to panic if a lock cannot be acquired; for example, if another thread is already holding the lock.

Summary

This chapter helps us build up on the concepts in *Chapter 3, Building a Custom Blockchain*. We've built out some core blockchain functionality, such as writing the code to work with our blockchain with functions for adding blocks and hashes and then writing the code for the server and the node to be able to serve our created blockchain. We are now a step closer to creating a blockchain.

In the next chapter, we will finish up our blockchain. We will write and break down the logic for transactions, wallets, memory pools, and UTXOs that complete a blockchain.

5

Finishing Up
Our Custom Blockchain

We started our journey of building a blockchain in *Chapter 3, Building a Custom Blockchain*, when we started outlining structs and functions. We then continued fleshing out those functions in *Chapter 4, Adding More Features to Our Custom Blockchain*. Now, in this chapter, we will take things forward and finish up our blockchain.

In the previous chapter, we built out some core blockchain functionality, such as writing the code to work with our blockchain with functions for adding blocks and hashes and then writing the code for the server and the node to be able to serve our created blockchain. Now, we will discuss the following topics:

- Adding memory pools
- Implementing transactions
- Utilizing **unspent transaction outputs (UTXOs)** and developing wallets
- Setting up configurations and utilities
- Understanding the Main.rs file
- Using your custom blockchain

Technical requirements

The code for each chapter is presented separately in different folders in the GitHub repository accompanying this book. For an uninterrupted learning journey, all the code pertaining to this chapter can be accessed in the dedicated GitHub repository. You can clone it with the following command:

```
git clone https://github.com/PacktPublishing/Rust-for-Blockchain-
Application-Development/.
```

This repository contains comprehensive code snippets, projects, and resources relevant to the material discussed. To better grasp the explanations, you are encouraged to clone the repository and explore its contents as we progress through the explanations.

> **An important point to note**
>
> While the divided code is in different folders with the name of the specific chapter, the combined code from *Chapters 3*, *4*, and *5* for the entire blockchain project is in a folder called `complete_blockchain`, so if you'd like to read the entire code and move between the functions for a better understanding, kindly refer to the code in this code file.

Adding memory pools

A **memory pool**, often referred to as a **mempool**, serves as a holding area for pending cryptocurrency transactions awaiting validation and inclusion in a block on a blockchain network. It stores unconfirmed transactions, acting as a temporary repository before miners select and verify them for block inclusion. Significantly impacting blockchain efficiency, the mempool plays a pivotal role in transaction processing and network performance. Its function involves validating transactions, ensuring they comply with network rules, and prioritizing them based on associated fees or other criteria. The mempool's role in decentralized systems, such as Bitcoin, Ethereum, and various altcoins, enhances transaction throughput and responsiveness. Applications of mempools extend beyond basic transaction storage; they influence fee estimation algorithms, network scalability solutions, and transaction acceleration services, which are vital in optimizing blockchain operations and user experience. Understanding the mempool's dynamics is crucial for developers, miners, and users navigating the intricacies of blockchain technology.

Our project will have a `memory_pool.rs` file, where the entire logic for memory pools is available. We will start with understanding a simple implementation of a memory pool data structure that allows for the storage, retrieval, and removal of transactions in a thread-safe manner. We will follow it up with a simple implementation of a data structure for tracking blocks that are in transit during a **peer-to-peer** (**P2P**) networking protocol.

Let's now look at the memory pool implementation in detail.

Implementing a memory pool

In the `memory_pool.rs` file, take a look at the following code block for `MemoryPool`:

```
pub struct MemoryPool {
    inner: RwLock<HashMap<String, Transaction>>,
}
```

The `MemoryPool` struct contains a field named `inner` of the `RwLock<HashMap<String, Transaction>>`type, which is a thread-safe read-write lock allowing multiple readers or a single writer at any given time to access a `HashMap` that maps `String` keys to `Transaction` values.

Let us now look at the definition of the `MemoryPool` structure in Rust, presenting a rudimentary implementation for managing pending transactions within a blockchain system. The `MemoryPool` struct contains essential functions to interact with and manipulate transactions stored within it. The definition will look like this:

```
impl MemoryPool {
    ...... .
}
```

Let us now understand one by one what all the functions inside the definition do:

- The `new()` function serves as a constructor for the `MemoryPool` struct. It initializes and returns a new instance of `MemoryPool`, setting up an internal `HashMap` wrapped within a `RwLock`. This `HashMap` is used to store transactions, ensuring safe concurrent access by employing a read-write lock mechanism:

  ```
  pub fn new() -> MemoryPool {
      MemoryPool {
          inner: RwLock::new(HashMap::new()),
      }
  }
  ```

- The `contains(&self, txid_hex: &str) -> bool` function checks whether a transaction with a specific transaction ID (`txid_hex`) exists within the memory pool. It does this by reading the internal `HashMap` within the `RwLock`, verifying whether the provided transaction ID is a key in the map. If the transaction ID exists in the pool, it returns `true`; otherwise, it returns `false`:

  ```
  pub fn contains(&self, txid_hex: &str) -> bool {
      self.inner.read().unwrap().contains_key(txid_hex)
  }
  ```

- The `add(&self, tx: Transaction)` function inserts a new transaction (`tx`) into the memory pool. It first generates a hexadecimal representation of the transaction ID using a hashing algorithm. Then, it acquires a write lock on the internal `HashMap` to ensure safe concurrent modification and inserts the transaction into the map using its transaction ID as the key:

  ```
  pub fn add(&self, tx: Transaction) {
      let txid_hex = HEXLOWER.encode(tx.get_id());
      self.inner.write().unwrap().insert(txid_hex, tx);
  }
  ```

- The get(&self, txid_hex: &str) -> Option<Transaction> function retrieves a transaction from the memory pool based on a given transaction ID (txid_hex). It reads the internal HashMap within the RwLock and attempts to find the transaction associated with the provided transaction ID. If the transaction exists in the pool, it returns a Some instance containing a clone of the transaction. If the transaction is not found, it returns None:

```
pub fn get(&self, txid_hex: &str) -> Option<Transaction> {
    if let Some(tx) = self.inner.read().unwrap().get(txid_hex) {
        return Some(tx.clone());
    }
    None
}
```

- The remove(&self, txid_hex: &str) function removes a transaction from the memory pool based on its transaction ID (txid_hex). It acquires a write lock on the internal HashMap and removes the transaction associated with the provided transaction ID, effectively deleting it from the pool. This function is to facilitate the auto removal of transactions that are rejected in mempools. Unconfirmed transactions reside only temporarily:

```
pub fn remove(&self, txid_hex: &str) {
    let mut inner = self.inner.write().unwrap();
    inner.remove(txid_hex);
}
```

- The get_all(&self) -> Vec<Transaction> function retrieves all transactions stored in the memory pool. It obtains a read lock on the internal HashMap and iterates through its entries, cloning each transaction into a vector (Vec). Finally, it returns this vector containing all transactions present in the memory pool:

```
pub fn get_all(&self) -> Vec<Transaction> {
    let inner = self.inner.read().unwrap();
    let mut txs = vec![];
    for (_, v) in inner.iter() {
        txs.push(v.clone());
    }
    return txs;
}
```

- The len(&self) -> usize function returns the current count of transactions in the memory pool. It acquires a read lock on the internal HashMap and retrieves the length of the map, representing the number of transactions currently stored in the memory pool:

```
pub fn len(&self) -> usize {
    self.inner.read().unwrap().len()
}
}
```

After understanding all the functions of the `MemoryPool` definition, let us move on to the `BlockInTransit` implementation.

The BlockinTransit implementation

Now, we will look at the code for the `BlockinTransit` implementation, which is part of the `memory_pool.rs` file itself.

This code will provide a simple implementation of a data structure for tracking blocks that are in transit during a P2P networking protocol. It allows for the storage, retrieval, and removal of blocks in a thread-safe manner.

Here's the code that you will see in the file:

```rust
pub struct BlockInTransit {
    inner: RwLock<Vec<Vec<u8>>>,
}
  impl BlockInTransit {
    pub fn new() -> BlockInTransit {
        BlockInTransit {
            inner: RwLock::new(vec![]),
    }
}
    pub fn add_blocks(&self, blocks: &[Vec<u8>]) {
        let mut inner = self.inner.write().unwrap();
        ...
}
    pub fn first(&self) -> Option<Vec<u8>> {
        ...
    }
    pub fn remove(&self, block_hash: &[u8]) {
        let mut inner = self.inner.write().unwrap();
        if let Some(idx) = inner.iter().position(|x| x.eq(block_hash))
{
            inner.remove(idx);
        }
    }
```

Take a look at *Figure 5.1*, which breaks down each function within the `BlockInTransit` structure:

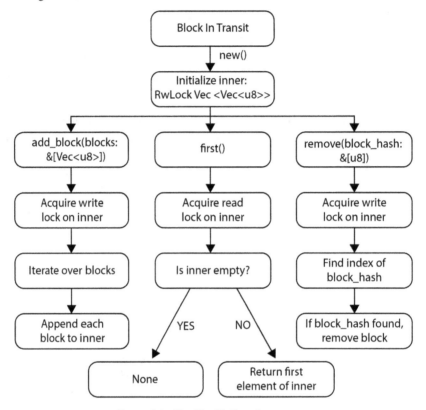

Figure 5.1 – The BlockInTransit structure

Let's talk about each of these functions in detail:

- `new()`: This function acts as a constructor for `BlockInTransit`, creating and returning a new instance. Inside `new()`, an empty `RwLock`-protected vector (`Vec`) is initialized to store blocks.

- `add_blocks(&self, blocks: &[Vec<u8>])`: The `add_blocks()` function allows the addition of multiple blocks to the `BlockInTransit` structure. It obtains a write lock on the internal vector to ensure exclusive modification and iterates through the provided blocks, converting each to a vector of bytes and appending it to the internal vector.

- `first(&self) -> Option<Vec<u8>>`: This function retrieves the first block stored in the `BlockInTransit` structure. It obtains a read lock on the internal vector and checks whether there is a block available. If present, it clones the first block as a vector of bytes and returns it wrapped in a `Some`. If the vector is empty, it returns `None`.

- `remove(&self, block_hash: &[u8])`: The `remove()` function deletes a specific block identified by its hash (`block_hash`) from `BlockInTransit`. It acquires a write lock, searches for the block's position in the vector using the provided hash, and removes it if found.

- `clear(&self)`: This function clears all blocks from `BlockInTransit`. It obtains a write lock and empties the internal vector, effectively removing all elements.

- `len(&self) -> usize`: The `len()` function returns the current number of blocks stored in `BlockInTransit`. It obtains a read lock, retrieves the length of the internal vector, and returns the count as a `usize` pointer.

These functions collectively enable the management of blocks within `BlockInTransit`, allowing for the addition, retrieval, removal, clearing, and count retrieval of blocks stored within the structure.

Now that we've delved into the functionalities of managing blocks within the `BlockInTransit` structure, let's pivot our focus toward a deeper exploration of the transactions occurring within our blockchain system.

Implementing transactions

Transactions in a blockchain network represent the transfer of assets or information between participants. In this section, we dissect the core components of transactions – TXInput, TXOutput, and the overarching `Transaction` struct. These structures intricately manage inputs, outputs, and transactional data, crucial for secure asset exchanges within the blockchain ecosystem.

All the code we write in this section will be available in the `transactions.rs` file.

Understanding TXInput transactions

We have already seen the `TXInput` struct in the previous chapter; now, let's understand the implementation functions for all `TXInput` transactions. It will look like the following code:

```
impl TXInput {

    pub fn new(txid: &[u8], vout: usize) -> TXInput {
        TXInput {
            txid: txid.to_vec(),
            vout,
            signature: vec![],
            pub_key: vec![],
        }
    }

    pub fn get_txid(&self) -> &[u8] {
        self.txid.as_slice()
```

```
    }

    pub fn get_vout(&self) -> usize {
        self.vout
    }

    pub fn get_pub_key(&self) -> &[u8] {
        self.pub_key.as_slice()
    }

    pub fn uses_key(&self, pub_key_hash: &[u8]) -> bool {
        let locking_hash = wallet::hash_pub_key(self.pub_key.as_
slice());
        return locking_hash.eq(pub_key_hash);
    }
}
```

The TXInput constituent fields are shown in *Figure 5.2*:

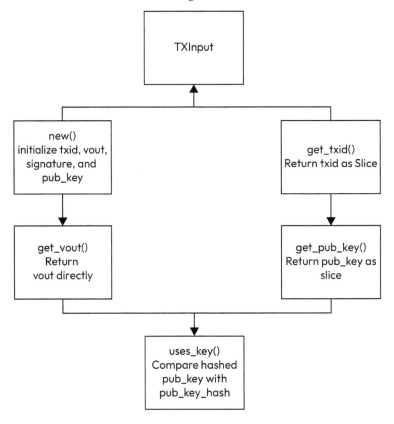

Figure 5.2 – TXInput fields

Here's a description of each of the fields:

- txid: A vector of bytes representing the ID of the transaction that created the output that this input is spending

- vout: An index that represents which output of the transaction with ID txid this input is spending

- signature: A vector of bytes that will contain a digital signature of the transaction that includes this input

- pub_key: A vector of bytes that will contain the public key of the owner of the funds being spent

The TXInput struct has the following methods:

- new: A constructor that creates a new TXInput instance with the specified txid and vout fields, an empty signature, and public key fields.

- get_txid: A method that returns a reference to the txid field of the input.

- get_vout: A method that returns the vout field of the input.

- get_pub_key: A method that returns a reference to the pub_key field of the input.

- uses_key: A method that takes a pub_key_hash byte vector as input and returns a Boolean indicating whether the pub_key field of the input corresponds to the specified pub_key_hash byte vector. This method calls a wallet::hash_pub_key() function to compute the hash of the public key and then compares it with the specified pub_key_hash byte vector.

This concludes our explanation of TXInput.

Let us now move on to TXOutput.

Understanding TXOutput transactions

The TXOutput struct manages transaction outputs within a blockchain, storing values and public key hashes. It facilitates the creation of new outputs, value retrieval, and verification of locked outputs using cryptographic hashes. This component plays a vital role in ensuring secure asset transfers and validation in blockchain transactions.

Here's the code that you will see in the file for the TXOutput struct:

```
impl TXOutput {
    pub fn new(value: i32, address: &str) -> TXOutput {
        ...
        return output;
    }

    pub fn get_value(&self) -> i32 {
```

```
            self.value
        }

    pub fn get_pub_key_hash(&self) -> &[u8] {
        self.pub_key_hash.as_slice()
    }

    fn lock(&mut self, address: &str) {
        ...
        self.pub_key_hash = pub_key_hash;
    }

    pub fn is_locked_with_key(&self, pub_key_hash: &[u8]) -> bool {
        self.pub_key_hash.eq(pub_key_hash)
    }
}
```

Let us now explore and understand the TXOutput struct, whose fields and methods are shown in *Figure 5.3*:

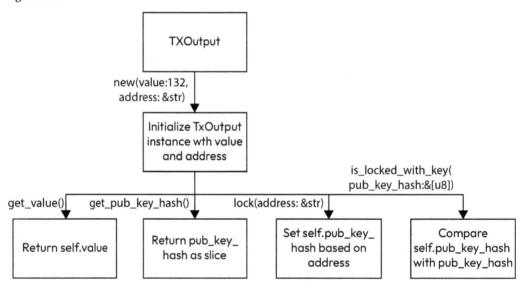

Figure 5.3 – TXOutput struct

Let's first discuss the fields:

- value: An integer that represents the value of the output
- pub_key_hash: A vector of bytes that represents the hash of the public key of the owner of the output

The TXOutput struct has the following methods:

- new: A constructor that creates a new TXOutput instance with the specified value and address, and sets the pub_key_hash field by calling the lock method

- get_value: A method that returns the value field of the output

- get_pub_key_hash: A method that returns a reference to the pub_key_hash field of the output

- lock: A private method that takes an address string as input, decodes it using the base58_ decode function, and sets the pub_key_hash field to the hash of the public key in the decoded payload

- is_locked_with_key: A method that takes a pub_key_hash byte vector as input and returns a Boolean indicating whether the pub_key_hash field of the output matches the specified pub_key_hash byte vector

We will now move on to the Transaction structure and explore and understand its functions and methods.

Understanding the Transaction implementation

This Transaction implementation manages transaction creation, validation, and signature verification in a blockchain. It constructs Coinbase and UTXO transactions, handles transaction signing and verification, and provides methods for serialization and deserialization of transaction data.

A Transaction struct contains inputs and outputs, where inputs refer to previous UTXOs that are being spent and outputs represent newly created UTXOs:

```
impl Transaction {
    pub fn new_coinbase_tx(to: &str) -> Transaction {
        let txout = TXOutput::new(SUBSIDY, to);
        let mut tx_input = TXInput::default();
        tx_input.signature = Uuid::new_v4().as_bytes().to_vec();
        ....
        return tx;
    }

    pub fn new_utxo_transaction(
        from: &str,
        to: &str,
        amount: i32,
        utxo_set: &UTXOSet,
    ) -> Transaction {
        ...
```

```
        let mut inputs = vec![];
        for (txid_hex, outs) in valid_outputs {
            ...
        }
        let mut outputs = vec![TXOutput::new(amount, to)];
        ...
        return tx;
    }
    ...
}
```

Figure 5.4 gives an overview of the `Transaction` implementation:

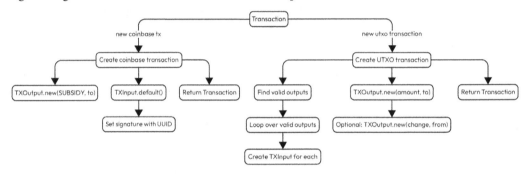

Figure 5.4 – Transaction implementation

Let's break down the functions within the `Transaction` implementation:

- `new_coinbase_tx(to: &str) -> Transaction`: This function creates a new Coinbase transaction, generating a transaction output (`txout`) with a specified value and recipient address. It constructs a transaction input (`tx_input`) with a unique signature using a UUID, combines these components into a `Transaction` struct, and computes its ID through hashing:

```
    pub fn new_coinbase_tx(to: &str) -> Transaction {
        let txout = TXOutput::new(SUBSIDY, to);
        let mut tx_input = TXInput::default();
        tx_input.signature = Uuid::new_v4().as_bytes().to_vec();
        ....
        return tx;
    }
```

- `new_utxo_transaction(from: &str, to: &str, amount: i32, utxo_set: &UTXOSet) -> Transaction`: This function constructs a new UTXO-based transaction by selecting spendable outputs and creating inputs for the transaction. It calculates the inputs required based on available outputs, manages outputs for the recipient and change, signs the transaction, and computes its ID through hashing:

```
pub fn new_utxo_transaction(
        from: &str,
        to: &str,
        amount: i32,
        utxo_set: &UTXOSet,
    ) -> Transaction {
        ...
        let mut inputs = vec![];
        for (txid_hex, outs) in valid_outputs {
            ...
        }
        let mut outputs = vec![TXOutput::new(amount, to)];
        ...
        return tx;
    }
```

- `trimmed_copy(&self) -> Transaction`: This internal function creates a trimmed copy of the transaction, excluding signatures, enabling signature verification without modifying the original transaction:

```
fn trimmed_copy(&self) -> Transaction {
        ...
    }
```

- `sign(&mut self, blockchain: &Blockchain, pkcs8: &[u8])`: The sign function signs the transaction inputs using the **Elliptic Curve Digital Signature Algorithm** (**ECDSA**). It retrieves previous transactions, prepares a copy for signature verification, signs inputs with the corresponding private keys, and updates the transaction with signatures:

```
fn sign(&mut self, blockchain: &Blockchain, pkcs8: &[u8]) {
        let mut tx_copy = self.trimmed_copy();
        for (idx, vin) in self.vin.iter_mut().enumerate() {
            ....
        }
    }
```

- `verify(&self, blockchain: &Blockchain) -> bool`: This function verifies transaction signatures against corresponding public keys. It checks for Coinbase transactions, prepares a trimmed copy, validates signatures against public keys, and ensures the correctness of previous transactions before confirming the authenticity of signatures:

```
pub fn verify(&self, blockchain: &Blockchain) -> bool {
        . . .
        let mut tx_copy = self.trimmed_copy();
        for (idx, vin) in self.vin.iter().enumerate() {
            let prev_tx_option = blockchain.find_
transaction(vin.get_txid());
            let prev_tx = prev_tx_option.unwrap();
            . . .
            let verify = crate::ecdsa_p256_sha256_sign_verify(
                . . .
            );
            if !verify {
                return false;
            }
        }
        true
    }
```

- `is_coinbase(&self) -> bool`: The `is_coinbase` function checks whether the transaction is a Coinbase transaction, verifying whether there's only one input and the public key length is 0:

```
pub fn is_coinbase(&self) -> bool {
        return self.vin.len() == 1 && self.vin[0].pub_key.len()
== 0;
    }
```

- `hash(&mut self) -> Vec<u8>`: This function generates the transaction's hash by creating a copy without the ID, serializing it, and computing its SHA-256 digest:

```
fn hash(&mut self) -> Vec<u8> {
        . . .
        crate::sha256_digest(tx_copy.serialize().as_slice())
    }
```

- **Other helper functions**: Additional functions such as `get_id`, `get_id_bytes`, `get_vin`, `get_vout`, `serialize`, and `deserialize` provide various utilities to retrieve transaction information, serialize the transaction to bytes, and deserialize it back from bytes:

```
pub fn get_id(&self) -> &[u8] {
        self.id.as_slice()
```

```
        }
        pub fn get_id_bytes(&self) -> Vec<u8> {
            self.id.clone()
        }
        pub fn get_vin(&self) -> &[TXInput] {
            self.vin.as_slice()
        }
        pub fn get_vout(&self) -> &[TXOutput] {
            self.vout.as_slice()
        }
        pub fn serialize(&self) -> Vec<u8> {
            bincode::serialize(self).unwrap().to_vec()
        }
        pub fn deserialize(bytes: &[u8]) -> Transaction {
            bincode::deserialize(bytes).unwrap()
        }
```

These functions collectively manage the creation, signing, verification, and manipulation of transactions within a blockchain network, ensuring their validity and security throughout the transaction life cycle.

Let us now understand the implementation of UTXOs and wallets.

Utilizing UTXOs and developing wallets

In this section, we will explore the implementation of UTXO and **wallet** functionalities. UTXO tracking and wallet management play pivotal roles in blockchain systems, ensuring transaction outputs' integrity and secure asset storage and transfer.

Let us first look at the UTXOSet structure.

Implementing UTXOSet

The UTXOSet structure helps manage UTXOs within a blockchain. It facilitates functionalities such as finding spendable outputs, reindexing outputs, updating outputs after block confirmation, and counting transactions within the blockchain.

You will find the following code in the file:

```
pub struct UTXOSet {
    blockchain: Blockchain,
}

impl UTXOSet {
  pub fn new(blockchain: Blockchain) -> UTXOSet {
```

```
            UTXOSet { blockchain }
      }

      pub fn get_blockchain(&self) -> &Blockchain {
          &self.blockchain
      }
pub fn find_spendable_outputs(...) -> (i32, HashMap<String,
Vec<usize>>) {
          let mut unspent_outputs: HashMap<String, Vec<usize>> =
HashMap::new();            (accmulated, unspent_outputs)
...
}
pub fn find_utxo(&self, pub_key_hash: &[u8]) -> Vec<TXOutput> {
          ...
      }
      pub fn count_transactions(&self) -> i32 {
          ...
      }
      pub fn reindex(&self) {
          ...
      }
      pub fn update(&self, block: &Block) {
          ...
      }

}
```

Let's break down the functionalities within the `UTXOSet` structure:

- `new(blockchain: Blockchain) -> UTXOSet`: This function initializes a new `UTXOSet` instance with a reference to the blockchain it operates on.

- `get_blockchain(&self) -> &Blockchain`: This retrieves a reference to the blockchain associated with the `UTXOSet` structure.

- `find_spendable_outputs(&self, pub_key_hash: &[u8], amount: i32) -> (i32, HashMap<String, Vec<usize>>)`: This method identifies spendable outputs for a given public key hash and required amount. It iterates through UTXOs, checks ownership, accumulates values, and forms a `HashMap` of transaction IDs to output indices for spendable outputs.

- `find_utxo(&self, pub_key_hash: &[u8]) -> Vec<TXOutput>`: This finds all UTXOs associated with a provided public key hash, iterating through the UTXO tree and gathering matching outputs.

- `count_transactions(&self) -> i32`: This function counts the number of transactions in the UTXO tree by iterating through its entries.

- `reindex(&self)`: This reindexes the UTXO tree by clearing it and rebuilding it from the blockchain's transaction outputs, ensuring consistency and correctness.

- `update(&self, block: &Block)`: This method updates the UTXO set after a block confirmation. It removes spent outputs and adds new outputs generated by transactions within the block, maintaining the integrity of the UTXO tree in the blockchain.

These functions collectively manage UTXOs within the blockchain, enabling the identification of spendable outputs, retrieval of UTXOs for specific public key hashes, updating the UTXO set after block confirmation, and maintaining the integrity of transaction outputs in the blockchain system.

Moving on, let us explore the implementation of wallets.

Implementing wallets

Let us define functionalities for creating and managing wallet addresses in a blockchain system. This includes methods for generating addresses, retrieving public keys, validating addresses, and converting public key hashes to wallet addresses:

```
impl Wallets {
    pub fn new() -> Wallets {
        ...
    }

    pub fn create_wallet(&mut self) -> String {
        ...
    }

    pub fn get_addresses(&self) -> Vec<String> {
        ...
    }

    pub fn get_wallet(&self, address: &str) -> Option<&Wallet> {
        ...
    }

    pub fn load_from_file(&mut self) {
        ...
    }

    fn save_to_file(&self) {
        ...
    }
}
```

Here is a breakdown of the code for the wallet and its functions:

- `new() -> Wallets`: Generates a new `Wallet` instance. It creates a new cryptographic key pair, extracts the public key, and initializes a `Wallets` structure with the generated keys:

```
pub fn new() -> Wallets {
        let mut wallets = Wallets {
                wallets: HashMap::new(),
        };
        wallets.load_from_file();
        return wallets;
}
```

- `get_address() -> String`: Constructs an address from the `Wallets` structure's public key. It combines a version number, a hashed version of the public key, and a checksum before encoding it in Base58 format, resulting in a unique address representation:

```
pub fn get_addresses(&self) -> Vec<String> {
        let mut addresses = vec![];
        for (address, _) in &self.wallets {
            addresses.push(address.clone())
        }
        return addresses;
}
```

- `get_public_key() -> &[u8]`: Retrieves the raw bytes representing the public key associated with the `Wallets` structure.

- `get_pkcs8() -> &[u8]`: Retrieves the raw bytes of the PKCS #8 representation of the private key associated with the `Wallets` structure.

- `hash_pub_key(pub_key: &[u8]) -> Vec<u8>`: Hashes the given public key using SHA-256 and then RIPEMD-160 hash functions to produce a hash representation:

```
pub fn hash_pub_key(pub_key: &[u8]) -> Vec<u8> {
    let pub_key_sha256 = crate::sha256_digest(pub_key);
    crate::ripemd160_digest(pub_key_sha256.as_slice())
}
```

- `checksum(payload: &[u8]) -> Vec<u8>`: Generates a checksum for a payload by applying a double SHA256 hash and extracting the first bytes, resulting in a verification code:

```
fn checksum(payload: &[u8]) -> Vec<u8> {
    let first_sha = crate::sha256_digest(payload);
    let second_sha = crate::sha256_digest(first_sha.as_slice());
    second_sha[0..ADDRESS_CHECK_SUM_LEN].to_vec()
}
```

- `validate_address(address: &str) -> bool`: Validates the integrity of an address by decoding it, separating its components, and recomputing the checksum. Returns `true` if the address is valid; otherwise, it returns `false`:

```
pub fn validate_address(address: &str) -> bool {
    let payload = crate::base58_decode(address);
    let actual_checksum = payload[payload.len() - ADDRESS_CHECK_
SUM_LEN..].to_vec();
    let version = payload[0];
    let pub_key_hash = payload[1..payload.len() - ADDRESS_CHECK_
SUM_LEN].to_vec();

    let mut target_vec = vec![];
    target_vec.push(version);
    target_vec.extend(pub_key_hash);
    let target_checksum = checksum(target_vec.as_slice());
    actual_checksum.eq(target_checksum.as_slice())
}
```

- `convert_address(pub_hash_key: &[u8]) -> String`: Converts a public key hash into an encoded address by appending a version number, the public key hash, and a checksum, then encoding it using Base58 encoding:

```
pub fn convert_address(pub_hash_key: &[u8]) -> String {
    let mut payload: Vec<u8> = vec![];
    payload.push(VERSION);
    payload.extend(pub_hash_key);
    let checksum = checksum(payload.as_slice());
    payload.extend(checksum.as_slice());
    crate::base58_encode(payload.as_slice())
}
```

These functions together allow for the creation, extraction, validation, and conversion of addresses associated with a cryptographic wallet, enabling secure transactions and **identity management** (**IdM**) within a blockchain system.

Let us shift our focus to developing wallets.

Wallets

Let us manage a collection of wallets within a blockchain application. Here's a breakdown of the `Wallets` struct and its associated methods.

The `Wallets` struct is organized like this:

```
impl Wallets{
...
}
```

Within this struct, we have the following code:

```
pub fn new() -> Wallets {
    let mut wallets = Wallets {
        wallets: HashMap::new(),
    };
    wallets.load_from_file();
    wallets
}
```

Let's break this down:

- `wallets: HashMap<String, Wallet>`: Represents a collection of wallets stored in a HashMap, where each wallet is associated with a unique address:

```
pub fn create_wallet(&mut self) -> String {
    let wallet = Wallet::new();
    let address = wallet.get_address();
    self.wallets.insert(address.clone(), wallet);
    self.save_to_file();
    address
}

pub fn get_addresses(&self) -> Vec<String> {
    self.wallets.keys().cloned().collect()
}

pub fn get_wallet(&self, address: &str) -> Option<&Wallet> {
    self.wallets.get(address)
}

pub fn load_from_file(&mut self) {
    // ...
}

fn save_to_file(&self) {
    // ...
}
}
```

- `wallets: HashMap<String, Wallet>`: Represents a collection of wallets stored in a HashMap, where each wallet is associated with a unique address.

- `new() -> Wallets`: Initializes a new instance of `Wallets`, creates an empty `HashMap` to store wallets, and attempts to load wallets from a file.

- `create_wallet() -> String`: Generates a new wallet, retrieves its unique address, inserts the wallet into the `wallets HashMap` using the address as the key, and saves the updated wallet collection to a file. Returns the newly created wallet's address.

- `get_addresses() -> Vec<String>`: Retrieves all addresses associated with the wallets stored in the `wallets HashMap` and returns them as a vector of strings.

- `get_wallet(&self, address: &str) -> Option<&Wallet>`: Retrieves a reference to a wallet by its address from the `wallets HashMap`. Returns `Some(wallet)` if the wallet exists; otherwise, it returns `None`.

- `load_from_file(&mut self)`: Attempts to load wallet data from a file. If the file exists, it reads the data and deserializes it into the `wallets HashMap`.

- `save_to_file(&self)`: Saves the contents of the `wallets HashMap` into a file. It serializes the wallets and writes the serialized data to the specified file.

The preceding code manages the creation, retrieval, storage, and persistence of wallets within the blockchain application, ensuring secure wallet management and persistence across sessions.

Moving on, let us explore configurations and utilities.

Setting up configurations and utilities

Setting up configurations and utilities within the code base establishes fundamental structures and utility functions pivotal for blockchain operations. These components encapsulate key configurations, cryptographic functions, and essential utilities indispensable for secure and efficient blockchain functionalities. Configurations encompass system configurations, including node addresses and mining specifics, ensuring the adaptability and customization of the blockchain environment. On the other hand, utilities house crucial cryptographic functions such as hashing algorithms, key generation, encoding, and signature verification, serving as foundational tools for secure data handling and authentication within the blockchain network. This section forms the backbone of system configurations and cryptographic operations essential for maintaining the integrity and security of blockchain-based systems.

Let us first look at the `Config` implementation.

The Config implementation

The `Config` structure serves as a centralized repository for managing configurations within a blockchain system. It encapsulates crucial information such as node addresses, mining settings, and related system configurations:

```
pub static GLOBAL_CONFIG: Lazy<Config> = Lazy::new(|| Config::new());

static DEFAULT_NODE_ADDR: &str = "127.0.0.1:2001";
const NODE_ADDRESS_KEY: &str = "NODE_ADDRESS";
const MINING_ADDRESS_KEY: &str = "MINING_ADDRESS";

pub struct Config {
    inner: RwLock<HashMap<String, String>>,
}
impl Config {
    pub fn new() -> Config {
        ...
    }

    pub fn get_node_addr(&self) -> String {
        ...
    }

    pub fn set_mining_addr(&self, addr: String) {
        ...
    }

    pub fn get_mining_addr(&self) -> Option<String> {
        ...
    }

    pub fn is_miner(&self) -> bool {
        ...
    }
}
```

The `new()` function initializes a new instance of `Config`. It reads the node address from the system environment variables, allowing for dynamic customization. If a specific node address is defined in the environment variables (`NODE_ADDRESS`), it overrides the default node address (`DEFAULT_NODE_ADDR`) to ensure flexibility in node setup. It then creates an internal `HashMap` holding the node address under the `NODE_ADDRESS_KEY` key.

The get_node_addr() method retrieves the node address stored within Config. It grants read access to the internal RwLock, allowing concurrent read operations. It retrieves the node address by accessing the value associated with NODE_ADDRESS_KEY in the HashMap, cloning the value to return it as a String value.

The set_mining_addr() function updates the mining address within the Config struct. It acquires a write lock to the internal RwLock, ensuring exclusive access for modification. The method inserts or updates the mining address (addr) under MINING_ADDRESS_KEY within the HashMap.

The get_mining_addr() method retrieves the mining address stored in Config. It grants read access to the internal RwLock, similar to get_node_addr(). It attempts to retrieve the mining address associated with MINING_ADDRESS_KEY from the HashMap. If found, it clones and returns the address as an Option<String> value. If absent, it returns None.

The is_miner() function checks whether a mining address is present in the Config struct. It acquires read access to the internal RwLock, permitting concurrent read operations. It verifies the existence of the mining address key (MINING_ADDRESS_KEY) within the HashMap, returning a Boolean value based on its presence.

Overall, this Config structure allows for the management and retrieval of critical system configurations, such as node addresses and mining settings, offering flexibility and customization options within a blockchain-based system.

Utility functions

Let's move on to a breakdown of each utility function:

```
pub mod utils;
use utils::base58_decode;
use utils::base58_encode;
use utils::current_timestamp;
use utils::ecdsa_p256_sha256_sign_digest;
use utils::ecdsa_p256_sha256_sign_verify;
use utils::new_key_pair;
use utils::ripemd160_digest;
use utils::sha256_digest;
```

Here's an overview of each function:

- current_timestamp(): This function retrieves the current timestamp as an integer representing milliseconds since the Unix epoch. It uses SystemTime::now() to capture the current time, calculates the duration since the Unix epoch, and converts it to milliseconds. It returns the timestamp as an i64 value.

- `sha256_digest(data: &[u8]) -> Vec<u8>`: This function performs a SHA-256 hash operation on the provided `data` input, returning the resulting hash as a vector of bytes. It initializes a hashing context with SHA-256, updates the context with the input data, generates the hash digest, and converts it to a vector of bytes for output.

- `ripemd160_digest(data: &[u8]) -> Vec<u8>`: This function calculates the `RIPEMD-160` hash of the input, returning the resulting hash as a vector of bytes. It creates a `RIPEMD-160` hasher, inputs the data, collects the resulting hash into a byte vector, and returns it.

- `base58_encode(data: &[u8]) -> String`: This function encodes the given byte slice using the Base58 encoding scheme and returns the encoded string representation. It utilizes `bs58 crate` to perform the encoding and converts the byte data into a Base58-encoded string.

- `base58_decode(data: &str) -> Vec<u8>`: Conversely, this function decodes a Base58-encoded string back to its original byte representation. It uses the `bs58 crate` to decode the input string and returns the decoded byte vector.

- `new_key_pair() -> Vec<u8>`: This function generates a new ECDSA key pair and returns the private key as a byte vector. It utilizes `EcdsaKeyPair` and `SystemRandom` from the `ring` crate to generate a private key in PKCS#8 format and converts it to a byte vector.

- `ecdsa_p256_sha256_sign_digest(pkcs8: &[u8], message: &[u8]) -> Vec<u8>`: This function signs the provided `message` parameter using the ECDSA `P-256 SHA-256` algorithm. Given a private key in PKCS#8 format (`pkcs8`), it creates an ECDSA key pair, signs the message, and returns the resulting signature as a byte vector.

- `ecdsa_p256_sha256_sign_verify(public_key: &[u8], signature: &[u8], message: &[u8]) -> bool`: This function verifies an ECDSA `P-256 SHA-256` signature against a provided `message` parameter using the corresponding `public_key` value. It constructs an unparsed public key from the `public_key` byte slice and uses it to verify the provided `signature` against the `message` parameter, returning a Boolean indicating the signature's validity.

These functions offer utilities for cryptographic operations (hashing, encoding, decoding, key generation, signing, and verification) commonly used in blockchain systems, ensuring secure and reliable transaction handling and data manipulation within the blockchain infrastructure.

Understanding the lib.rs file

This file orchestrates the blockchain system by structuring its components into modules, which we have discussed in the previous sections:

```
blockchain_rust/
├── src/
│   ├── block.rs            // Block module
│   ├── blockchain.rs       // Blockchain module
```

```
|   ├── config.rs          // Blockchain module
|   ├── lib.rs        // Blockchain module
|   ├── main.rs        // Blockchain module
|   ├── memory_pool.rs        // Blockchain module
|   ├── node.rs          // Node module
|   ├── req_functions.rs            // Node module
|   ├── server.rs        // Server module
|   ├── structs.rs        // Server module
|   ├── transactions.rs      // Transaction module
|   ├── utils.rs        // Transaction module
|   ├── utxo_set.rs        // UTXO Set module
|   ├── wallets.rs        // Wallets module
└── Cargo.toml            // Rust package configuration
```

Let's break this down:

- **Block and Blockchain**: `Block` and `Blockchain` modules manage the blocks and the chain of blocks, respectively, forming the foundational structure of the blockchain

- **UTXOSet**: The `UTXOSet` module handles UTXOs, vital for verifying transactions' validity

- **ProofOfWork**: Within the `ProofOfWork` module, the mining process is defined using the **proof-of-work** (**PoW**) consensus mechanism to add blocks to the chain securely

- **Transaction**: The `Transaction` module encapsulates transaction-related functionalities, facilitating value transfer between participants

- **Wallet and Wallets**: `Wallet` manages individual user wallets, while `Wallets` oversees multiple wallets, facilitating user interaction with the blockchain system

- **Server and Node**: The `Server` module handles communication between nodes, and the `Node` module manages the network's nodes and their interactions

- **MemoryPool**: `MemoryPool` keeps track of pending transactions awaiting confirmation by inclusion in a block

- **Configuration**: The `Config` module manages system configurations, including node addresses and mining settings

- **Utilities**: The `utils` module houses utility functions essential for cryptographic operations, encoding, decoding, hashing, timestamp management, and cryptographic key generation, used across various blockchain functionalities

This file's structural organization streamlines development, promoting modularity, code reusability, and easy maintenance of the blockchain system. Each module encapsulates specific functionalities, facilitating their integration, testing, and management within the broader blockchain architecture. Additionally, the **separation of concerns** (**SoC**) among modules enhances code readability and scalability while supporting the robustness and integrity of the blockchain implementation.

We will now move our focus to one of the most important files, the Main.rs file.

Understanding the Main.rs file

The Main.rs file serves as the entry point and orchestrator of operations in this Rust blockchain application. It defines **command-line interface (CLI)** options using the StructOpt crate, enabling users to interact with various functionalities of the blockchain system. Each command represents a specific action, from creating a blockchain and managing wallets to sending transactions and exploring the blockchain's state. The code structure, coupled with the utilization of modules and external crates, forms a cohesive ecosystem to execute blockchain operations efficiently. The code is organized around the Opt structure, which encapsulates different commands specified under the Command enum. These commands offer diverse functionalities such as creating a new blockchain, managing wallets, checking balances, sending transactions, examining the blockchain's state, and initializing nodes.

Let us get into an explanation of the code in the Main.rs file:

```
struct Opt {
    #[structopt(subcommand)]
    command: Command,
}
```

Let's break this down:

- const MINE_TRUE: usize = 1;: Defines a constant flag representing a value for triggering immediate mining when adding a new block. The MINE_TRUE flag is used within the Command::Send functionality to control block mining.

- struct Opt: This structure is used in conjunction with the StructOpt derive macro to define command-line options. It encapsulates various commands under the Command enum, facilitating different interactions with the blockchain system.

- enum Command: Represents a set of distinct commands accessible via the CLI. Each command offers specific functionalities:

```
enum Command {
    #[structopt(name = "createblockchain", about = "Create a new
blockchain")]
    Createblockchain { address: String },
    #[structopt(name = "createwallet", about = "Create a new
wallet")]
    Createwallet,

    ...
    #[structopt(name = "send", about = "Add new block to
chain")]
    Send { from: String, to: String, amount: i32, mine: usize },
    #[structopt(name = "printchain", about = "Print blockchain
```

```
all block")]
    Printchain,
    ...
}
```

- **Createblockchain**: Allows the creation of a new blockchain with a provided address to receive the genesis block reward.

- **Createwallet**: Facilitates the creation of a new wallet address within the blockchain system.

- **GetBalance**: Checks and displays the balance of a specified wallet address.

- **ListAddresses**: Lists all wallet addresses available locally.

- **Send**: Enables the addition of a new block to the blockchain by initiating a transaction between two addresses, with an option to immediately mine the block.

- **Printchain**: Prints details of the entire blockchain, including block hashes, timestamps, and transaction information.

- **Reindexutxo**: Rebuilds the UTXO index set.

- **StartNode**: Initializes a node for interacting with the blockchain, optionally enabling mining mode.

- **fn main()**: The main entry point of the application, where commands provided through the CLI are processed and executed. The function matches the provided commands against predefined options and triggers the corresponding actions:

```
fn main() {
    env_logger::builder().filter_level(LevelFilter::Info).
init();
    let opt = Opt::from_args();
    match opt.command {
        Command::Createblockchain { address } => {
            ...
        }
        Command::Createwallet => {
            ...
        }
        ...
        Command::Send { from, to, amount, mine } => {
            ...
        }
        Command::Printchain => {
            ...
        }
        ...
```

```
            Command::StartNode { miner } => {
                ...
            }
        }
```

- `Command::Createblockchain`: Initiates the creation of a new blockchain and reindexes the UTXO set.

- `Command::Createwallet`: Generates a new wallet address.

- `Command::GetBalance`: Checks the balance of a specified wallet address by querying the UTXO set.

- `Command::ListAddresses`: Lists all available wallet addresses.

- `Command::Send`: Sends a transaction between addresses, optionally mining a new block immediately.

- `Command::Printchain`: Iterates through the blockchain, displaying block details and transaction information.

- `Command::Reindexutxo`: Rebuilds the UTXO set index.

- `Command::StartNode`: Initializes a node, allowing interaction with the blockchain and server initialization for networking.

Each command in the `Command` enum corresponds to distinct functionalities, facilitating user interaction with the blockchain system via a CLI. The `fn main()` function serves as a comprehensive orchestrator, executing specific actions based on user inputs, ensuring the effective operation of the blockchain application.

The `Main.rs` file acts as the control center, enabling users to interact with the blockchain system via a CLI. It seamlessly processes user inputs, triggering specific actions and interactions with the blockchain components, such as blocks, transactions, wallets, and network nodes. The modular and structured approach facilitates diverse blockchain functionalities, enhancing user accessibility and system usability within this Rust-based blockchain application.

Next up, we will see how we can use the blockchain that we have created across the three chapters.

Using your custom blockchain

Before running your blockchain, ensure your environment is set up correctly. This step involves installing Rust, which is necessary for compiling and managing the project, given its development in Rust language. Here's what you need to do:

1. **Install Rust**: Visit the official Rust website and follow the instructions to install Rust and Cargo on your system. Cargo is Rust's package manager and build system.

2. **Compile the project**: The next step is to compile the project, which you can do as follows:

I.   Open a terminal and navigate to the Rust-Advanced-Blockchain directory within the project.

II. Compile the project using Cargo. This will download dependencies and compile the project:

```
cargo build -release
```

III. The --release flag builds the project in release mode, which optimizes the binary for performance.

After running the preceding command, you will see the following result in the terminal:

```
blockchain_rust 0.1.0
USAGE:
    blockchain_rust.exe <SUBCOMMAND>
FLAGS:
    -h, --help          Prints help information
    -V, --version       Prints version information
SUBCOMMANDS:
    createblockchain    Create a new blockchain
    createwallet        Create a new wallet
    getbalance          Get the wallet balance of the target address
    help                Prints this message or the help of the given
subcommand(s)
    listaddresses       Print local wallet addres
    printchain          Print blockchain all block
    reindexutxo         rebuild UTXO index set
    send                Add new block to chain
    startnode           Start a node
```

Let us test out each command one at a time.

Creating a new blockchain

Initialize a new blockchain, creating a genesis block and setting up the initial state of the blockchain. The command that we will use is createblockchain:

```
blockchain_rust.exe createblockchain --address YOUR_ADDRESS
```

Here's an example using a specific address:

```
blockchain_rust.exe createblockchain --address 1AVsR...Xb3N
```

This command initializes a new blockchain. The --address flag specifies the address that receives the reward for the genesis block, which is the first block in the blockchain. You then need to do the following:

1. **Generate a wallet address**: Before you can create a blockchain, you need a wallet address to receive the genesis block reward. Use the createwallet command to generate a new wallet and address.

2. **Execute the command**: Replace YOUR_ADDRESS with your wallet address. This address will receive the genesis block reward, effectively initializing your blockchain with its first block.

3. **Blockchain initialization**: The command creates a genesis block and sets up the blockchain's initial state, marking the beginning of your blockchain network.

4. **Outcome**: Upon successful execution, your blockchain is initialized, and the specified address receives the genesis block reward. This step is foundational, as it prepares the blockchain for subsequent transactions and blocks.

After initializing the blockchain, you will want to perform additional actions such as creating wallets, checking balances, and making transactions. Each action corresponds to a specific command, which we will explore in detail, following the preceding format to ensure clarity and ease of understanding.

Creating a new wallet

Generate a new wallet, including a private and public key pair, with the public key serving as your blockchain address for receiving and sending currency. The command that we will use is createwallet:

```
blockchain_rust.exe createwallet
```

For example, imagine you've just launched the CLI and entered the preceding command.

Upon execution, the system might display something like this:

```
Your new wallet address: 1AVsR...Xb3N
```

The createwallet command is designed to generate a new set of cryptographic keys (private and public keys). The public key is transformed into a wallet address, which you'll use to receive and send blockchain assets. Follow these steps:

1. **Launch the command**: Simply type blockchain_rust.exe createwallet into your CLI. No additional arguments are needed for this command.

2. **Key generation**: Behind the scenes, the blockchain application generates a secure private key, which is essential for signing transactions and proving ownership of assets. The corresponding public key is then derived from the private key.

3. **Address creation**: The public key undergoes a series of cryptographic transformations to produce a unique wallet address. This address can be shared publicly to receive funds.

4. **Safety reminder**: It's crucial to keep your private key secure. Anyone with access to your private key can control your blockchain assets associated with its corresponding wallet address.

You now have a unique wallet address associated with a new set of keys. This address is your identity on the blockchain for receiving and sending currency. It's the first step toward interacting with the blockchain ecosystem, allowing you to participate in transactions.

Wallets are fundamental to blockchain technology, serving as the interface through which users interact with the blockchain. They store your private keys securely and provide tools to sign transactions, proving ownership of blockchain assets without revealing the private keys themselves. Understanding how to create and manage your wallet is crucial for anyone looking to actively engage with blockchain networks, whether for sending/receiving currency, participating in **decentralized applications** (**dApps**), or managing digital assets.

In the next section, we will explore how to check the balance of a wallet address using the `getbalance` command, offering insight into managing and monitoring your blockchain assets effectively.

Checking the wallet balance

Determine the balance of a specific wallet address, allowing you to see how much currency you have in the blockchain network, using the `getbalance` command:

```
blockchain_rust.exe getbalance --address YOUR_WALLET_ADDRESS
```

For example, after obtaining your wallet address by creating a wallet, you can check its balance as follows:

```
blockchain_rust.exe getbalance --address 1AVsR...Xb3N
```

The command line might return something like this:

```
Balance of 1AkXxhYCrtiYzV1eEAtEbGQDr6qjddCfq5: 10
```

The `getbalance` command retrieves the total amount of currency (for example, BTC) available at a specified wallet address. This includes the sum of all UTXOs linked to the address. Follow the next steps:

1. **Identify the address**: Use the address of the wallet for which you want to check the balance. This address was generated when you created a new wallet.

2. **Execute the command**: Replace YOUR_WALLET_ADDRESS with your actual wallet address. The system then scans the blockchain for all transactions involving this address, calculating the balance by summing all UTXOs that you can spend.

3. **Understanding your balance**: The displayed balance reflects the amount of currency you can spend. It's essential for managing your assets and planning transactions.

You've successfully checked the balance of your wallet address, gaining insights into your available blockchain assets. This step is crucial for effective asset management and strategic transaction planning within the blockchain ecosystem.

Wallet balances are pivotal in blockchain ecosystems, providing a clear view of your available assets for transactions. Understanding your balance is key to executing transactions, as it ensures you have enough currency for the intended operations, including transaction fees that may apply. Checking your balance regularly helps in monitoring incoming transactions and confirming that your funds are correctly updated after sending currency.

Next, we'll explore the `startnode` command, focusing on how to set up and start a node in the blockchain network, marking a significant step toward engaging with the blockchain more dynamically, such as participating in network consensus or hosting a copy of the blockchain data.

Starting a node

Initiate a node within the blockchain network, enabling it to connect to other nodes, participate in the network's consensus mechanism, and contribute to the blockchain's overall functionality, using the `startnode` command:

```
blockchain_rust.exe startnode --address YOUR_WALLET_ADDRESS
```

For instance, to start a node and participate in the blockchain network, you might use the following command:

```
blockchain_rust.exe startnode --address 1AVsR...Xb3N
```

The system might indicate the successful launch of the node with messages such as this:

```
Starting node 1AVsR...Xb3N
Node successfully connected to the blockchain network.
```

The `startnode` command is used to launch a new node in the blockchain network. The node will attempt to connect to existing nodes, synchronize blockchain data, and participate in processing transactions and blocks. Let's look at the process in more detail:

1. **Node identification**: The `--address` flag specifies the wallet address associated with the node. This address can be used for receiving rewards if the node participates in consensus mechanisms such as mining or validating transactions, depending on the blockchain's protocol.

2. **Networking**: Upon starting, the node begins to connect to the network, discovering other nodes and exchanging blockchain data to ensure it has the latest state of the blockchain.

3. **Active participation**: The node now contributes to the network's security and integrity by validating transactions, proposing new blocks (if applicable), and relaying information.

Your node is now an active participant in the blockchain network, contributing to its operation and security. Starting a node is a significant step toward deeper engagement with the blockchain, offering opportunities to contribute to transaction validation, block creation, and the overall governance of the network.

With your node up and running, you're well positioned to perform transactions within the network, such as sending currency to other addresses. This capability is crucial for leveraging the blockchain for its intended purpose: secure, transparent, and decentralized value exchange.

In the next section, we'll detail the send command, which allows you to execute transactions, transferring value between addresses on the blockchain. This step is essential for engaging in economic activities facilitated by the blockchain, whether for personal transactions, investment, or as part of dApps.

Sending currency

Perform a transaction to transfer currency from one wallet to another, illustrating the process of sending blockchain assets, by using the send command:

```
blockchain_rust.exe send --from SENDER_ADDRESS --to RECEIVER_ADDRESS
--amount AMOUNT --mine
```

To send an amount of 10 from your address to another address, you would execute the following command:

```
blockchain_rust.exe send --from 1AVsR...Xb3N --to 1BvBM...SExy
--amount 10 --mine
```

You may get an output like this:

```
Sending 10 from 1AVsR...Xb3N to 1BvBM...SExy
Transaction successfully added to the block.
```

The send command is used to transfer currency from one address to another. This is one of the core functionalities of any blockchain, enabling the movement of assets across the network. Let's take a look at the process:

- **Transaction details**: Specify the sender and receiver addresses, along with the amount of currency to be transferred. The --mine flag immediately mines a new block containing this transaction on your node, confirming the transaction without waiting for another miner to do so.

- **Transaction execution**: The blockchain application validates the transaction, ensuring the sender has sufficient balance and the transaction details are correct. It then broadcasts the transaction to the network for confirmation.

- **Confirmation and mining**: If the --mine option is used, your node will attempt to mine a new block containing this transaction. Once mined, the transaction is confirmed, and the receiver's balance is updated.

You've successfully sent currency from one address to another, demonstrating the ability to execute and confirm transactions within the blockchain network. This step is fundamental to blockchain technology, showcasing how assets are transferred and managed securely and transparently.

Next in the sequence is the `listaddresses` command, which we haven't discussed yet. Let's detail this command next, providing you with a guide on how to list all the wallet addresses you've generated or used on your local node.

Listing all wallet addresses

Let's see how to display all wallet addresses that have been generated or used by the user on the local node, helping to keep track of all possible transaction endpoints. When you want to see all the wallet addresses associated with your node, simply run the `listaddresses` command:

```
blockchain_rust.exe listaddresses
```

The command might output a list like this:

```
1. 1AVsR...Xb3N
2. 1BvBM...SExy
3. 1C1bC...Df3Y
...
```

This command provides a comprehensive list of all the wallet addresses that the user's node can recognize. It's particularly useful for managing multiple addresses or verifying which addresses are available for transactions. Let's take a look at the process:

- **Simple execution**: There are no additional arguments required. The command scans the local node's data, compiling a list of all known addresses.

- **Reviewing addresses**: The output helps users identify which addresses they can use for receiving and sending transactions. It can also aid in managing assets across multiple addresses.

Users can easily view and manage their wallet addresses, enhancing their ability to engage with the blockchain network effectively.

After listing all the wallet addresses, we'll move on to the `printchain` command, which allows users to view the entire blockchain.

Printing the blockchain

View all blocks in the blockchain from the most recent back to the genesis block, providing a comprehensive look at the entire chain's history, by using the `printchain` command:

```
blockchain_rust.exe printchain
```

This command typically generates a detailed list of all blocks in the blockchain, presented in reverse order (starting with the most recent). For each block, you might see information such as the block's hash, the hash of the previous block, timestamps, and the list of transactions contained within:

```
Block Hash: 0000a1b2c3d4e...
Prev. Hash: 0000987z6y5xw...
Timestamp: 2023-01-01 12:00:00
Transactions:
    Tx Hash: 123456abcdef...
    Value: 10 BTC
    From: 1AVsR...Xb3N
    To: 1BvBM...SExy
...
```

The `printchain` command is invaluable for anyone looking to verify transactions or understand the flow of currency within the blockchain. It offers transparency into the blockchain's operation, showing how blocks are interconnected and how the blockchain grows over time. It has the following advantages:

- **Historical insight**: By printing the blockchain, users gain insights into the sequence of transactions and blocks, which is essential for auditing and verifying the integrity of the blockchain

- **Verification and trust**: This command reinforces the blockchain's transparency and trustworthiness, allowing users to independently verify the existence and details of transactions

You have a clear and detailed view of the blockchain's history, enhancing your understanding of its integrity and the sequence of transactions that have occurred over time.

With the blockchain's history made transparent through the `printchain` command, users can ensure the accuracy and integrity of their transactions and the blockchain as a whole. This level of insight is crucial for trust in blockchain systems, allowing for independent verification of all activities on the chain.

Next, we'll discuss the `reindexutxo` command, which is vital for maintaining an accurate and efficient representation of available transaction outputs within the blockchain.

Rebuilding the UTXO set

Reconstruct the index of UTXOs, ensuring the blockchain accurately reflects the current set of spendable transaction outputs, by using the `reindexutxo` command:

```
blockchain_rust.exe reindexutxo
```

Upon completion, the system might not display detailed information but internally updates the UTXO set. You may see a simple confirmation message like this:

```
UTXO set successfully reindexed.
```

The `reindexutxo` command is crucial for maintaining the integrity and efficiency of the blockchain. The UTXO set represents all UTXOs available in the blockchain, which is essential for verifying transactions and calculating wallet balances. Here's why the process is important:

- **Necessity of reindexing**: Over time, transactions are added to the blockchain, and UTXOs are spent and created. This command ensures that the UTXO set is current, reflecting only those outputs that are truly unspent.

- **Enhancing performance**: By keeping the UTXO set accurate and minimal, this process helps optimize transaction verification and balance calculation, ensuring the blockchain remains efficient and scalable.

- **Maintaining blockchain integrity**: Regularly reindexing the UTXO set can prevent discrepancies in transaction processing and balance calculations, contributing to the overall health and accuracy of the blockchain.

The blockchain's UTXO set is now up to date, ensuring that all transactions and balances are accurately reflected based on the current state of the blockchain. This step is essential for the correct operation of the blockchain, particularly for transaction verification and wallet balance calculations.

Reindexing the UTXO set is a behind-the-scenes maintenance task that, while not directly visible to users, plays a crucial role in the blockchain's performance and reliability. This process ensures that the system can efficiently process transactions and accurately report wallet balances, reinforcing the trust and integrity of the blockchain network.

With the explanation of the `reindexutxo` command completed, we have covered essential commands for interacting with the blockchain, from starting a node and creating wallets to sending transactions and maintaining the blockchain's integrity. Each command plays a specific role in the ecosystem, enabling users to engage with the blockchain in a meaningful and effective way.

Having explored the foundational commands to interact with your custom blockchain, you're now equipped with the tools necessary to create, manage, and analyze a blockchain network. From initializing the blockchain with `createblockchain`, managing wallets through `createwallet`, conducting transactions with `send`, to ensuring the blockchain's integrity with `reindexutxo`, you've taken significant steps toward understanding the operational aspects of blockchain technology.

Summary

In this chapter, we embarked on finalizing the essential elements of a blockchain system, emphasizing the mechanisms that facilitate secure and efficient transactions. Initially, we focused on completing the blockchain code to ensure a solid foundational structure and protocol. We then delved into memory pools, which play a pivotal role in managing transaction queues before their incorporation into the blockchain. This discussion naturally progressed to understanding UTXOs, vital for processing and validating transactions and indicative of the spendable digital currency amount for users.

Subsequently, the chapter shifted its focus to wallet development, highlighting its significance not just in storing digital currency but also in securing transactions and maintaining user identity within the blockchain through private key management. We explored how blocks are finalized in the blockchain, detailing the selection, verification, and addition of transactions from the memory pool, a process crucial for the blockchain's integrity and continuity. Finally, we examined methods and protocols for transferring value among users, concluding the chapter by demonstrating how these integrated concepts enable a secure and functional blockchain system.

In the next chapter we will shift our focus to building smart contracts using the Foundry framework on an Ethereum blockchain.

Part 3: Building Apps

In this part, we will build decentralized applications that can run on popular chains such as Ethereum, Solana, and NEAR.

This part has the following chapters:

- *Chapter 6, Using Foundry to Build on Ethereum*
- *Chapter 7, Exploring Solana by Building a dApp*
- *Chapter 8, Exploring NEAR by Building a dApp*

6

Using Foundry to Build on Ethereum

The world of blockchain technology continues to evolve rapidly, pushing the boundaries of **decentralized applications (dApps)** and **smart contracts**. **Ethereum**, with its robust ecosystem and vibrant community, remains at the forefront of this revolution. Within the Ethereum network, smart contracts serve as the backbone for executing secure and transparent agreements without the need for intermediaries. Ethereum is the most popular and widely used blockchain for the development of smart contracts, and hence is the first blockchain that we will learn to work with in this book.

This chapter delves into the innovative **Foundry framework**, an advanced development tool designed to streamline the process of building smart contracts on the Ethereum blockchain. The Foundry framework is a powerful and flexible solution that empowers developers to create efficient and reliable smart contracts, facilitating the growth and adoption of blockchain-based applications.

Furthermore, we will delve into the practical aspects of using the Foundry framework, providing step-by-step tutorials and code examples to illustrate its implementation. From setting up the development environment to deploying and testing smart contracts, this chapter will guide you through the entire development lifecycle using the Foundry framework.

By the end of this chapter, you will have a comprehensive understanding of the Foundry framework and its capabilities, empowering you to build robust and efficient smart contracts on the Ethereum blockchain.

In this chapter, we will cover the following topics:

- Introducing Ethereum and Foundry
- Exploring Foundry
- Understanding Foundry with **Cast**, **Anvil**, and **Chisel**
- Testing and deployment
- A project using Foundry

Introducing Ethereum and Foundry

In this section, we will explore Ethereum's capabilities and dive into the Foundry framework, a powerful tool for building smart contracts on the Ethereum blockchain.

We will discuss the key features and benefits of Foundry and provide step-by-step guidance on getting started with the framework. From setting up the development environment to deploying and testing smart contracts, this topic serves as a comprehensive resource for developers looking to harness the potential of Ethereum and leverage the Foundry framework to create efficient and reliable blockchain applications.

So, let's explore a bit further and learn about this framework, which is gaining significant popularity.

Understanding Ethereum

Ethereum is a decentralized, open source blockchain platform. Launched in 2015 by Vitalik Buterin, Ethereum introduced a significant innovation by extending the capabilities of blockchain technology beyond mere financial transactions.

At its core, Ethereum allows developers to create and deploy smart contracts, which are self-executing agreements with the terms of the contract directly written into code. This enables a wide range of applications, from **decentralized finance** (**DeFi**) and decentralized exchanges to supply chain management and many other dApps. Ethereum's smart contracts are written using **Solidity**, which is Turing-complete, meaning these contracts can execute any algorithm or computational task, making it a versatile platform for building complex decentralized applications.

> **Self-executing agreements**
> Self-executing agreements are the code that executes when certain conditions are met, like how agreements work in real life, but automatically.

Ethereum operates using its native cryptocurrency called **Ether** (**ETH**), which serves as a means of exchange and incentivizes miners to secure the network. Ethereum's consensus mechanism, known as **Proof of Stake** (**PoS**), has transitioned from the previous **Proof of Work** (**PoW**) system, reducing energy consumption and improving scalability.

The Ethereum ecosystem is highly active and vibrant, with a robust developer community continually creating new tools, frameworks, and libraries to enhance development on the platform. Ethereum's decentralized nature also fosters a multitude of projects and protocols built on top of it, including decentralized exchanges such as Uniswap, lending platforms such as Aave, and stablecoins such as DAI. Stablecoins are vital in Ethereum, a form of cryptocurrency tied to assets such as fiat or commodities. They offer stability amid crypto volatility. DAI, an Ethereum stablecoin, is pegged to the US dollar, ensuring steadier value for transactions and DeFi activities, and reducing susceptibility to price shifts seen in Bitcoin or Ethereum.

Interoperability is another essential feature of Ethereum, allowing developers to interact and exchange assets across different blockchain networks through standards such as ERC-20 (fungible tokens) and ERC-721 (**non-fungible tokens (NFTs)**). This interoperability contributes to the growth of the DeFi sector and the emergence of the metaverse.

Metaverse and blockchain

In the context of blockchain and cryptocurrencies, the metaverse has gained attention as a potential application area where users can own, trade, and interact with digital assets, such as NFTs representing virtual land, digital art, and virtual fashion. These digital assets can be bought, sold, and used across various virtual worlds or platforms, creating an interconnected digital universe. As the metaverse continues to evolve, blockchain standards such as ERC-20 and ERC-721 play a role in enabling seamless asset exchange and ownership within these virtual environments.

Why Rust and Foundry?

When it comes to building dApps for the Ethereum blockchain, choosing the right programming language and development framework is crucial. In recent years, **Rust** has emerged as an excellent option, known for its performance, safety, and expressive syntax. When combined with the Foundry framework, which is built using Rust, developers can unlock a powerful and efficient toolkit for creating dApps on the Ethereum blockchain.

Rust stands out as a language that prioritizes memory safety and concurrency without sacrificing performance. It achieves this through its ownership model and strict compiler checks, preventing common pitfalls such as null pointer dereferences and data races. These features make Rust an ideal choice for building secure and reliable smart contracts, where the accuracy and integrity of the code are paramount. Moreover, Rust's focus on performance ensures that dApps built with it can handle complex transactions and scale effectively.

The Foundry framework takes advantage of Rust's strengths and provides a comprehensive set of tools and libraries for Ethereum development. It offers an abstraction layer that simplifies the process of building and interacting with smart contracts, reducing the complexity and boilerplate code typically associated with Ethereum development. Foundry provides a higher-level interface for contract deployment, contract interaction, event handling, and transaction management.

One of the notable features of the Foundry framework is its emphasis on security. By leveraging Rust's safety guarantees, the framework minimizes vulnerabilities and reduces the risk of exploits or attacks. This focus on security is crucial for dApps handling sensitive data, managing financial transactions, or dealing with user assets.

Additionally, the Foundry framework promotes modularity and code reusability. It encourages the use of small, composable contracts, allowing developers to build complex systems by combining and integrating these modular components. This modular approach enhances code maintainability and flexibility, making it easier to upgrade and adapt dApps over time.

Furthermore, the Rust ecosystem offers excellent support for Ethereum development, with libraries such as `ethers-rs` and `web3-rs` providing powerful tools for interacting with the Ethereum network. These libraries seamlessly integrate with the Foundry framework, enabling developers to leverage existing Rust resources and utilities to enhance their dApps.

This is the reason why, in this book, we have chosen Foundry as our framework for working with Ethereum.

It's important to note that Solidity, the programming language used for Ethereum smart contract development, is constantly evolving with new versions being released. When using Foundry or any framework in conjunction with Solidity, it's crucial to be aware of these updates. The versions of Solidity discussed in this book may become outdated over time. Therefore, always check the latest Solidity version on the official Solidity website at `https://solidity.io/` and ensure compatibility with the tools and frameworks you are using. This practice helps maintain the relevance and functionality of your smart contracts in the rapidly changing landscape of Ethereum development.

Installing Foundry

In order to get started with Foundry, we need to have it installed on our system. You can build the Foundry framework from its source on GitHub (`https://github.com/foundry-rs/foundry`), but the best way is to use a tool called **FoundryUp** that helps us manage the different versions of Foundry along with the updates. Along with this, it enables us to upgrade and downgrade based on the project requirements.

FoundryUp is a cross-platform tool and the same instructions work for Ubuntu, Windows, and macOS.

To install FoundryUp, you just need to start your terminal and run the following command:

```
curl -L https://foundry.paradigm.xyz | bash
```

Once the installation is complete, you can use the `foundryup` command. By running `foundryup` without any additional parameters, you will install the latest precompiled binaries for `forge`, `cast`, `anvil`, and `chisel` (which are included in Foundry). For more options, such as installing from a specific version or commit, refer to the `foundryup --help` command.

We can also build from source. To begin, it is essential to have the Rust compiler and **Cargo**, which is the Rust package manager. The most straightforward method of acquiring these tools is by utilizing `rustup.rs`.

For Windows users, you will need to install and use Git Bash or **Windows Subsystem for Linux** (**WSL**) as your terminal, since FoundryUp currently does not support PowerShell or cmd. Also, another requirement is a recent edition of Visual Studio, installed with the **Desktop Development With C++ Workloads** option.

You can either run the `foundryup` command or the following commands, which you need to enter:

```
foundryup --branch master
foundryup --path path/to/foundry
```

There's a single command that you could enter to get the same result:

```
cargo install --git https://github.com/foundry-rs/foundry --profile
local --force foundry-cli anvil chisel
```

Alternatively, you have the option of manually building Foundry from a local copy of the repository:

```
# clone the repository
git clone https://github.com/foundry-rs/foundry.git
cd foundry
# install Forge + Cast
cargo install --path ./cli --profile local --bins --force
# install Anvil
cargo install --path ./anvil --profile local --force
# install Chisel
cargo install --path ./chisel --profile local –force
```

Additionally, Foundry offers the capability to operate entirely within a Docker container. If Docker is not currently installed on your system, you can follow the instructions on how to install Docker at `https://docs.docker.com/engine/install/`.

Once Docker is installed, you can retrieve the most recent release by executing the following command:

```
docker pull ghcr.io/foundry-rs/foundry:latest
```

You can also build the Docker image locally from the `foundry` repository by just running the following command:

```
docker build -t foundry .
```

Now we can get to work by leveraging the capabilities of Foundry to streamline our development process.

First steps with Foundry

In the previous section, we installed Foundry, and now that we have it, we can try out a few commands to explore it.

Forge is a command-line tool that you get with Foundry and you can use it to carry out various operations. We will learn a lot more about Forge in the *Overview of Forge* section, but for now, we're going to use Forge to explore our newly installed Foundry in this section.

The `init` command that we get in the `forge` CLI helps us initialize a new foundry project, and that's the command we will now run:

1. Let's first start a new project with Foundry using the following command:

    ```
    $ forge init hello_foundry
    ```

2. After this command, forge generates four different directories for us, namely, `lib`, `script`, `src`, and `test`.

3. After you have changed into the project root directory, which was generated with the previous command, it can be built with this command:

    ```
    $ forge build
    ```

4. We can also run tests on this project by using the following command:

    ```
    $ forge test
    ```

Once all these commands work, we can be sure that Foundry was correctly installed.

In the next section, we will dive deeper into Foundry and understand some important concepts.

Exploring Foundry

By now, we know quite a bit about Ethereum and Foundry. We also understand the various benefits of using Foundry. In the previous section, we learned how to start a new Foundry project using the `forge init` command.

In this section, we will cover more details about Foundry, such as how to work on an existing Foundry project, how to manage dependencies. We will learn in detail about Forge – the primary component of Foundry and also the Forge CLI tool, which makes it possible for us to work with Foundry projects.

Working on an existing Foundry project

In many cases, you might join a team as a new engineer or developer on an existing project and you might need to contribute to this project. With Foundry, it is super simple to join a new project and start contributing.

Getting started with an existing project using the Foundry framework for Ethereum is a seamless process that allows developers to quickly dive into the development journey. The Foundry framework offers a straightforward and intuitive approach, making it easy for developers to get up and running in no time.

There are four main steps involved that you need to remember: importing the project, installing dependencies, building the project, and then testing the project. Forge makes it easy to do all of these, so let's get started.

If you have obtained an existing project that utilizes Foundry, getting started is a breeze. Follow these simple steps:

1. Download the project from a suitable source. For instance, let's consider cloning the template repository from GitHub:

    ```
    $ git clone https://github.com/abigger87/template
    $ cd template
    ```

2. Install your dependencies. Run `forge install` to install the submodule dependencies required by the project. This command ensures that all necessary components are in place for smooth development.

    Once the project is set up, you can proceed with building and testing it using Foundry's intuitive commands.

3. Utilize `forge build` to initiate the build process. This command compiles the project and prepares it for deployment, ensuring all dependencies are resolved and the code is ready for execution.

4. Ensure the reliability and functionality of your project by executing `forge test`. This command runs the suite of tests designed for your project, validating the behavior and performance of the smart contracts, and verifying that they meet the expected requirements.

With these steps completed, you are all set to start working on the existing project using the Foundry framework. From here, you can leverage Foundry's comprehensive set of tools and commands to build, compile, test, and deploy your Ethereum smart contracts with ease.

Dependencies

The Foundry framework for Ethereum provides a streamlined approach to managing dependencies, making it easy for developers to incorporate external libraries and packages into their smart contract projects.

Whether you need to add new dependencies, remap existing ones, update them to newer versions, or remove unnecessary dependencies, Foundry offers a seamless workflow.

Additionally, Foundry ensures compatibility with popular Ethereum development tools such as **Hardhat**, enhancing flexibility and convenience.

Adding new dependencies

To add new dependencies to your project, follow these steps:

1. Open the `Forge.toml` file in your project directory.

2. Locate the `[dependencies]` section.

3. Add a new line for each dependency in the format `"<dependency-name> = <version>"`.

4. Save the changes.

5. Foundry will automatically download and include the specified dependencies in your project when using `forge install` or `forge build`.

6. Another way to add new dependencies is to use the `forge install <name_of_ dependancy>` command, which directly downloads and installs the required dependency.

Remapping dependencies

In certain cases, you might need to remap existing dependencies to different versions or repositories. To remap a dependency, follow these steps:

1. Open the `Forge.toml` file in your project directory.

2. Locate the `[remappings]` section.

3. Add a new line in the format `"<dependency-name> = <new-source>"`.

4. Save the changes.

5. Foundry will update the dependency resolution to use the specified new source when using `forge install` or `forge build`.

Updating dependencies

To update dependencies to newer versions, use the following command:

```
$ forge update
```

Foundry will check for any available updates for your project's dependencies and automatically update them to the latest compatible versions.

Deleting dependencies

To remove unnecessary dependencies from your project, follow these steps:

1. Open the `Forge.toml` file in your project directory.

2. Locate the `[dependencies]` section.

3. Remove the line corresponding to the dependency you wish to delete.

4. Save the changes.

5. Foundry will remove the specified dependency from your project when using `forge install` or `forge build`.

Compatibility with Hardhat

Hardhat is one of the popular development environments that aid in simplifying the process of creating dApps, which you can read more about here: `https://hardhat.org/`. It can help with everything, including compiling, testing, deploying, and debugging smart contracts. Foundry is designed to seamlessly integrate with Hardhat. By utilizing Foundry with Hardhat, developers can leverage the benefits of both frameworks in harmony.

To ensure compatibility, follow these steps:

1. Set up your project with Hardhat using the desired configuration.

2. Follow the steps mentioned earlier to add, remap, update, or delete dependencies in your Foundry project.

3. Use the respective commands of each framework within your project's directory.

The compatibility between Foundry and Hardhat allows developers to enjoy the robustness and flexibility of both frameworks, combining their capabilities to create and manage Ethereum smart contract projects efficiently.

Project layout

Understanding the project structure is crucial for maintaining a clean and structured code base. Here is an overview of the typical project structure in the Foundry framework:

- **Contracts directory**: The contracts directory is where you store your smart contract files. It contains the Solidity (`.sol`) files that define your contracts, their functionality, and their interactions.

- **Test directory**: The `test` directory holds the test files for your smart contracts. It contains Solidity test files or JavaScript files that define tests using frameworks such as Mocha or Hardhat.

- **Scripts directory**: The `scripts` directory is where you can include any auxiliary scripts or deployment scripts that are required for your project. These scripts can be written in JavaScript or any other supported scripting language.

- **Configuration files**: Foundry uses configuration files to manage project-specific settings. The most important configuration file is `foundry.toml`, which contains project-level settings and dependencies.

- **Lib directory**: The `lib` directory is where you can include external library dependencies. These libraries can be added as Git submodules or manually included in the `lib` folder.

- **Build directory**: The `build` directory is automatically generated by Foundry when you compile your smart contracts. It contains the compiled bytecode, **application binary interface (ABI)**, and other artifacts generated during the build process. This is named `out` by default when it is generated.

- **Dist directory**: The `dist` directory is where you can place any distribution files or artifacts that are produced during the deployment or release process. These files can include compiled contracts, documentation, or other project-specific deliverables.

By adhering to this project structure, developers can maintain a clear separation of concerns, organize their code base effectively, and ensure smooth collaboration with other team members. The structure also aligns with industry best practices, making it easier to understand and navigate the project code base. Overall, the project structure in the Foundry framework promotes consistency, modularity, and maintainability in Ethereum smart contract development projects.

Overview of Forge

Forge is a crucial component of the Foundry framework, designed to enhance the development experience and empower developers to build Ethereum smart contracts with efficiency and ease. As an integral part of the Foundry ecosystem, Forge provides a comprehensive set of tools, libraries, and utilities that streamline the smart contract development lifecycle.

One of the primary objectives of Forge is to simplify the process of writing, deploying, and interacting with smart contracts on the Ethereum blockchain. It achieves this by abstracting away the complexities and intricacies of the underlying **Ethereum virtual machine (EVM)** and providing a high-level interface for developers. This abstraction layer allows developers to focus on the business logic of their smart contracts rather than getting caught up in low-level implementation details.

Forge offers a user-friendly development environment that facilitates smooth contract deployment. It provides utilities for compiling smart contract source code into bytecode, generating contract artifacts, and handling contract deployment to the Ethereum network. With Forge, developers can easily manage the lifecycle of their contracts, including versioning, upgrades, and interacting with existing contract instances.

Furthermore, Forge simplifies contract interaction by providing intuitive APIs and utilities. Developers can seamlessly interact with their deployed contracts, query contract state, and execute contract functions using a clean and straightforward syntax. Forge takes care of the underlying web3 interactions, making it easier to integrate Ethereum functionality into dApps built on the Foundry framework.

Security is a paramount concern when it comes to smart contract development, and Forge prioritizes it by offering built-in security features. It encourages best practices for secure coding, such as input validation and protection against re-entrancy attacks. Forge also incorporates advanced testing utilities, allowing developers to write comprehensive unit tests and ensure the reliability and robustness of their smart contracts.

Additionally, Forge promotes modularity and code reusability. It encourages developers to break down their smart contracts into smaller, composable components, which can be easily integrated and reused in different contracts. This modular approach simplifies code maintenance and upgrades and also facilitates collaboration between developers working on different parts of a contract system.

Forge Standard Library overview

Forge Standard Library (**Forge Std**) is a set of contracts that provide essential functionality for writing tests, such as the following:

- Shortcuts (or cheat codes, as people like to call them) to modify blockchain state for easy testing of contracts.

- Hardhat-style logging functionality.

- Basic utilities for Solidity scripting.

- **DSTest** is a testing library within the Foundry framework, used primarily for Ethereum smart contract development. It provides an assortment of assertion functions for validating contract behavior, alongside utilities for setting up test conditions and monitoring gas usage. It is enhanced by a superset with standard libraries, cheat codes, and Hardhat console integration.

There are several benefits to using Forge Std, including the following:

- It makes writing tests easier and faster

- It provides a consistent and well-tested set of functionalities

- It is compatible with both Foundry and Hardhat

Forge Std currently comprises six standard libraries that enhance various aspects of smart contract development:

- **Std Logs**: Std Logs builds upon the logging events available in the `DSTest` library, providing expanded functionality for logging purposes.

- **Std Assertions**: Std Assertions expands upon the assertion functions found in the DSTest library, offering additional capabilities for validating contract behavior.

- **Std Cheats**: Std Cheats are wrappers around Forge cheat codes, ensuring their safe usage and enhancing **developer experience** (**DX**). You can easily access Std Cheats by invoking them as internal functions within your test contract.

- **Std Errors**: Std Errors provides wrappers for common internal Solidity errors and reverts, simplifying error handling. They are especially useful when combined with the `expectRevert` cheat code, eliminating the need to remember internal Solidity panic codes. Accessing Std Errors is done through `stdError`, as it is a library.

- **Std Storage**: Std Storage simplifies the manipulation of contract storage by facilitating the identification and modification of storage slots associated with specific variables. The test contract includes a Std Storage instance called `stdstore`, through which you can access the functionalities. Remember to add using the `stdStorage` function for std Storage in your test contract.

- **Std Math**: Std Math is a library that provides useful mathematical functions not natively available in Solidity. Accessing these functions is done through `stdMath`, as it is a library. For example, to obtain the absolute value of `-10`, you can utilize `stdMath.abs(-10)` to retrieve the desired result.

Forge commands

Forge provides developers with a powerful set of commands for building, testing, and deploying Ethereum smart contracts. These commands streamline the development process and enhance efficiency. Here are some of the most important Forge commands:

- `forge new <project-name>`: This command creates a new project, generating the necessary directory structure and configuration files for your Ethereum smart contract development.

- `forge compile`: This compiles the smart contracts within your project, generating bytecode and ABI artifacts. This command ensures that your contracts are syntactically correct and prepares them for deployment.

- `forge test`: This executes the unit tests written for your smart contracts. It runs a suite of tests to validate the behavior and functionality of your contracts, ensuring they perform as expected.

- `forge deploy`: This deploys your smart contracts to the Ethereum blockchain. This command takes care of interacting with the network, handling the deployment process, and providing you with the contract address for further interaction.

- `forge upgrade <contract-name>`: This upgrades an existing contract to a new version. This command enables you to introduce improvements, bug fixes, or additional features to your contract while preserving the contract's state and data.

- `forge interact <contract-name>`: This launches an interactive shell to interact with deployed smart contracts. It provides a convenient way to invoke contract functions, query states, and handle events emitted by the contract.

- `forge verify <contract-name>`: This verifies your smart contract's source code on blockchain explorers such as **Etherscan**. This command adds transparency and ensures that others can review and validate your contract's code.

- `forge export`: This generates a client library or artifact that can be used by external applications to interact with your smart contracts. It simplifies the integration process for other developers or systems that need to interact with your contracts.

These Forge commands, among others available in the Foundry framework, equip developers with a comprehensive toolkit for Ethereum smart contract development.

Understanding Foundry with Cast, Anvil, and Chisel

In this section, we will delve into the three key components of the Foundry framework: **Anvil**, **Cast**, and **Chisel**.

These powerful tools play a pivotal role in enhancing the development experience and efficiency when building smart contracts on the Ethereum blockchain. Anvil provides a robust testing framework, Cast offers a comprehensive library for contract deployment and management, and Chisel facilitates smart contract upgradeability.

Together, these components form the backbone of the Foundry framework, empowering you to write, test, deploy, and upgrade Ethereum smart contracts with ease and confidence.

Overview of Cast

Cast is a fundamental component of the Foundry framework, specifically designed to simplify the testing process for Ethereum smart contracts. You can look more into cast at the following link: `https://book.getfoundry.sh/cast/`. As an integral part of the Foundry ecosystem, Cast provides you with a comprehensive and intuitive testing framework to ensure the reliability, security, and proper functioning of your smart contracts.

The primary objective of Cast is to streamline the testing process by offering a range of utilities and functionalities tailored for Ethereum smart contract testing. It provides you with a user-friendly interface to write comprehensive unit tests, integration tests, and functional tests for smart contracts. By leveraging Cast, you can verify the correctness of your contract logic and detect potential vulnerabilities or bugs before deploying the contracts to the Ethereum network.

With Cast, you can simulate different scenarios and test various aspects of their smart contracts. You can design test cases to cover edge cases, exceptional conditions, and boundary conditions to validate the behavior and responses of the contracts under different circumstances. Cast provides utilities for mocking external dependencies, simulating transactions, and managing test accounts, ensuring that the testing environment accurately represents real-world conditions.

One of the key features of Cast is its ability to generate and manage test data. It simplifies the process of creating test scenarios by providing utilities for generating random or specific input data for contract functions. This functionality enables you to thoroughly test your contracts with a wide range of input variations and ensure that the contract's behavior remains consistent and predictable.

Cast also integrates seamlessly with other components of the Foundry framework, such as Forge and the development environment. This integration enables developers to write tests that interact with their contracts directly, leveraging the high-level APIs provided by Forge. This not only simplifies the testing process but also promotes code reusability and consistency across the development lifecycle.

Overview of Anvil

Anvil is designed to enhance the deployment and management process of Ethereum smart contracts. You can read more about Anvil at the following link: `https://book.getfoundry.sh/anvil/`. As an integral part of the Foundry ecosystem, Anvil provides you with a robust and user-friendly set of tools and utilities for deploying, upgrading, and managing smart contracts on the Ethereum blockchain.

The primary objective of Anvil is to simplify and streamline the deployment process of smart contracts. It provides you with a high-level interface that abstracts away the complexities of interacting with the Ethereum network. With Anvil, you can easily compile your smart contract source code, generate bytecode, and deploy your contracts to the Ethereum network without getting bogged down in the intricacies of low-level deployment procedures.

Anvil also offers comprehensive contract management capabilities within the Foundry framework, enabling you to seamlessly handle contract upgrades and versioning. With Anvil, you can create upgradeable smart contracts, allowing for the seamless introduction of new features or bug fixes without disrupting existing contract functionality or requiring data migration. This ability to upgrade contracts simplifies maintenance and ensures that the contract remains up to date and adaptable.

Furthermore, Anvil provides utilities for contract verification and interaction. You can easily verify your smart contracts on Etherscan or other blockchain explorers, ensuring transparency and trustworthiness. Anvil also offers an intuitive API for interacting with deployed contracts, simplifying the process of invoking contract functions, querying contract states, and handling events emitted by the contracts.

Anvil's integration with other components of the Foundry framework, such as Forge and Cast, further enhances its capabilities. This integration allows you to deploy, manage, and test your smart contracts with ease, ensuring a cohesive and efficient development process.

Overview of Chisel

Chisel is designed to enhance Ethereum smart contract development through code generation capabilities. You can learn more about Chisel at the following link: `https://book.getfoundry.sh/chisel/`. As an integral part of the Foundry ecosystem, Chisel provides developers with a versatile toolkit for generating smart contract code, improving efficiency and productivity in the development process.

Chisel simplifies and accelerates the creation of Ethereum smart contracts by automating code generation. It offers you the ability to define contract templates and utilize code generation techniques to quickly generate boilerplate code for various contract components. This significantly reduces the amount of manual coding required, streamlining the development process and promoting code consistency.

Chisel provides a high-level interface for defining contract templates and specifying the desired contract structure, including functions, events, and modifiers. With Chisel, developers can easily define reusable templates for common contract patterns, such as ERC-20 tokens or decentralized exchanges, and generate fully functional contract code with minimal effort.

The code generation capabilities of Chisel also promote best practices and adherence to standards. By generating code based on predefined templates, Chisel ensures that contracts adhere to industry-standard patterns, reducing the likelihood of errors or vulnerabilities. This approach improves the overall quality, security, and maintainability of smart contracts built with the Foundry framework.

Furthermore, Chisel seamlessly integrates with other components of the Foundry framework, such as Forge and Anvil. This integration enables you to combine the benefits of code generation with the deployment and management functionalities provided by other components. The generated code can be easily deployed, managed, and tested within the Foundry ecosystem, creating a cohesive and efficient development experience.

Cast, Anvil, and Chisel important commands

Cast, Anvil, and Chisel are instrumental in streamlining various aspects of smart contract development on the Ethereum blockchain. With Cast, you can easily deploy and manage contracts, Anvil empowers you to write comprehensive tests for your contracts, and Chisel enables contract upgradability.

By understanding and mastering these important commands, you will gain the necessary tools to efficiently develop, test, deploy, and upgrade Ethereum smart contracts using the Foundry framework.

Let's dive into the details and demonstrate how to utilize the power of Cast, Anvil, and Chisel commands to enhance your smart contract development workflow.

Cast commands

Some common cast commands are as follows:

- `cast new <contract-name>`: This creates a new contract file using the specified name. This command initializes a contract template with a basic structure that can be customized according to your requirements.

- `cast generate <contract-name>`: This generates contract-specific tests based on the defined templates. It automates the process of creating tests for the contracts, saving development time and effort.

- `cast upgrade <contract-name>`: This upgrades an existing contract to a new version. This command helps introduce improvements or bug fixes to an existing contract while preserving the contract's state and data.

Anvil commands

Let's cover some commonly used Anvil commands:

- `anvil deploy`: This deploys the smart contracts to the Ethereum blockchain. This command takes care of contract deployment, handling network interactions, and providing deployment information such as contract addresses and transaction receipts.

- `anvil interact <contract-name>`: This opens an interactive shell to interact with deployed contracts. It allows developers to invoke contract functions, query states, and listen to events emitted by the contracts in real time.

- `anvil test`: This executes the unit tests written for the smart contracts. This command runs a suite of tests to verify the correctness and functionality of the contracts, ensuring they perform as intended.

Chisel commands

These chisel commands are essential for high productivity:

- `chisel generate <contract-name>`: This generates contract code based on predefined templates. This command simplifies contract creation by automating code generation for various contract components, reducing manual coding effort, and promoting code consistency.

- `chisel validate`: This validates the contract code against best practices and standards. This command ensures that the generated code follows industry-recommended patterns, improving contract quality, security, and maintainability.

Let's now learn how to test and deploy the smart contracts.

Testing and deployment

In this section, we will explore various aspects of testing and deployment. You will gain a deep understanding of how to ensure the reliability, security, and efficiency of your smart contracts.

We will begin by delving into the realm of testing, where we will cover the essential techniques and tools offered by Foundry. We'll learn how to write robust tests for our smart contracts using the Anvil testing framework, and we'll explore different testing methodologies, including unit tests, integration tests, and functional tests. We will also discover how to leverage fork testing to simulate real-world scenarios, fuzz testing to identify vulnerabilities, and invariant and differential testing to validate the integrity of your contracts.

Next, we'll dive into the deployment process and uncover the power of the Cast component, which simplifies and streamlines the deployment and management of your smart contracts. We'll guide you through the steps of deploying contracts to both local and live Ethereum networks. You will learn how to verify the deployed contracts using tools such as Etherscan, ensuring transparency and trust in your deployments.

Additionally, we'll explore advanced features of the Foundry framework that enable you to gather critical insights into your smart contract deployments. You will also discover how to generate detailed gas reports to optimize your contract's efficiency and cost-effectiveness and learn how to capture and analyze snapshots of your contract's state, enabling you to track changes and debug potential issues.

Throughout this section, we'll provide hands-on examples and practical guidance, enabling you to apply the knowledge directly to your own smart contract projects. By mastering the testing and deployment techniques offered by the Foundry framework, you'll gain the confidence and expertise needed to build reliable and secure Ethereum smart contracts.

So, let's dive into the world of testing and deployment with Foundry, and unlock the full potential of your smart contract development journey.

Writing tests

For Ethereum smart contracts, tests are typically written in Solidity, the programming language for Ethereum smart contracts. In Solidity testing, the outcome of a test function determines whether it passes or fails. If the test function reverts (throws an exception), the test is considered a failure. However, if the function completes without any issues, the test is considered a success.

To facilitate the process of writing tests in Solidity, Forge Std provides a test contract that offers a comprehensive set of functionalities. This test contract is widely recommended for writing tests with Forge.

In this section, we will explore the fundamental approach to writing tests using the functions provided by Forge Std's test contract. This contract serves as an extension of DSTest, enhancing its capabilities. More advanced features of Forge Std will be covered in subsequent sections.

To access the functions offered by DSTest and Forge Std's test contract, you need to import the `Test.sol` file and inherit from the test contract in your own test contract. This provides you with access to basic logging and assertion functionalities, allowing you to perform assertions and log information during test execution.

By importing the test contract and inheriting from it, you can leverage its features to enhance your test cases and ensure the correctness and reliability of your Ethereum smart contracts. Forge Std's test contract simplifies the process of writing tests in Solidity and provides essential utilities for effective testing. Let's look at an example of a basic test:

```solidity
pragma solidity 0.8.10;
// The following command imports the Test.sol file:
import "forge-std/Test.sol";
contract ContractBTest is Test {
    uint256 testNumber;
    function setUp() public {
        testNumber = 42;
    }
    function test_NumberIs42() public {
        assertEq(testNumber, 42);
    }
    function testFail_Subtract43() public {
```

```
        testNumber -= 43;
    }
}
```

In Forge, tests are written using specific keywords that provide structure and define the behavior of the test cases. Here are the keywords commonly used in Forge tests:

- `setUp`: This is an optional function that is invoked before each test case is executed. It allows you to set up any necessary preconditions or initialize variables for the test. For example, look at the following code:

  ```
  function setUp() public {
      testNumber = 42;
  }
  ```

- `test`: Functions prefixed with `test` are considered individual test cases. These functions are executed as separate tests to validate specific behaviors or functionalities. For instance, see the following:

  ```
  function test_NumberIs42() public {
  assertEq(testNumber, 42);
  }
  ```

- `testFail`: The `testFail` prefix is used to indicate that the test case should fail. If the function does not revert (throw an exception), the test is considered a failure. This is useful for testing scenarios where failure is expected. For example, see the following:

  ```
  function testFail_Subtract43() public {
  testNumber -= 43;
  }
  ```

- **Revert testing**: A recommended practice is to use the `test_Revert[If|When]_Condition` pattern along with the `expectRevert` cheat code. This provides more detailed information about the specific revert condition and error. For instance, look at the following:

  ```
  function test_CannotSubtract43() public {
  vm.expectRevert(stdError.arithmeticError);
  testNumber -= 43;
  }
  ```

By utilizing the `expectRevert` cheat code in combination with a specific condition, such as an arithmetic error, you can accurately verify that the expected revert behavior occurs.

These keywords and conventions in Forge tests contribute to organized and structured testing. They enable you to define and execute individual test cases while ensuring that the expected behavior, including reverts and failures, is properly validated.

Fork and fuzz testing

Fork and fuzz testing are crucial components of the Foundry framework for Ethereum, enabling developers to strengthen the testing process and uncover potential vulnerabilities in smart contracts. Let's delve into these concepts with real-world examples:

- **Fork testing**: Fork testing involves creating a local replica of the Ethereum network to simulate real-world scenarios. This allows for comprehensive testing without interacting with the live network. In Foundry, you can leverage tools such as Ganache or Hardhat to set up a local fork. Here's an example:

```
function test_ForkSimulation() public {
    // Create a local fork of the Ethereum network
    Fork fork = new Fork();

    // Perform test operations on the forked network
    // ...

    // Assert the expected outcomes
    // ...
}
```

- **Fuzz testing**: Fuzz testing, also known as **fuzzing**, involves providing random or mutated inputs to a smart contract to identify potential vulnerabilities. Foundry integrates with the **Echidna** tool, designed specifically for Ethereum smart contract fuzzing. Here's an example:

```
function test_FuzzingContract() public {
    // Define properties and invariants for fuzz testing
    Property[] properties = [Property1, Property2, Property3];

    // Run fuzz tests with Echidna
    EchidnaTestRunner.run(properties);
}
```

By running fuzz tests with Echidna and specifying properties to test, you can automatically generate random inputs to uncover unexpected behaviors or vulnerabilities in your smart contracts.

The combination of fork testing and fuzz testing in the Foundry framework provides a robust approach to enhancing the security and reliability of Ethereum smart contracts. It allows developers to simulate real-world scenarios and identify potential vulnerabilities through randomized input generation. By leveraging these testing techniques, developers can mitigate risks and ensure the solidity of their Ethereum-based projects.

Invariant and differential testing

Invariant and **differential** testing are essential techniques in the Foundry framework for Ethereum, providing developers with powerful tools to enhance the testing process and bolster smart contract security. Let's explore these concepts with actual examples:

- **Invariant testing**: Invariant testing involves verifying specific properties that should remain constant throughout a smart contract's execution. For instance, let's consider a simple token contract. An invariant could be that the total supply of tokens should always equal the sum of individual balances. In Foundry, you can write a test case to validate this invariant:

  ```
  function test_TotalSupplyEqualsSumOfBalances() public {
      // Deploy and initialize the token contract
      MyToken token = new MyToken();
      token.transfer(address(1), 100);
      token.transfer(address(2), 200);

      // Validate the invariant
      assertEq(token.totalSupply(), token.balanceOf(address(1)) +
  token.balanceOf(address(2)));
  }
  ```

- **Differential testing**: Differential testing helps identify discrepancies or vulnerabilities by comparing the behavior of different implementations of the same contract. Consider a contract that implements a voting system. You can employ differential testing to compare two versions of the contract to ensure they produce the same results. Here's an example:

  ```
  function test_DifferentialVoting() public {
      // Deploy the old and new versions of the voting contract
      VotingContract oldVersion = new OldVotingContract();
      VotingContract newVersion = new NewVotingContract();

      // Execute identical inputs and compare results
      oldVersion.vote(1);
      newVersion.vote(1);

      assertEq(oldVersion.getResult(), newVersion.getResult());
  }
  ```

By performing invariant and differential testing within the Foundry framework, you can verify critical properties and ensure consistent behavior across different contract versions. These testing techniques contribute to the overall security, reliability, and correctness of Ethereum smart contracts.

Deployment and verification

Up until now, we've learned how Foundry makes it super simple for us to build smart contracts. But once the contracts are built, they can be deployed to a testnet or any of the EVM-compatible chains, including Ethereum, and Foundry makes the deployment process easier for us as well. You might also want to verify the deployment, and this can also be easily done with Foundry, so let's take a look:

- **Deployment**: Foundry simplifies the deployment process of smart contracts by providing utilities and abstractions. For example, using the deployment capabilities in the framework, you can deploy a contract and perform initialization steps easily. Here's an example:

```
function test_ContractDeployment() public {
    // Deploy the contract
    MyContract contract = new MyContract();

    // Perform initialization steps
    contract.initialize();

    // Assert the contract was deployed successfully
    assert(contract.isDeployed());
}
```

- **Verification**: Foundry supports the verification of smart contracts, which is crucial for establishing trust and transparency. By utilizing tools such as the **Hardhat Etherscan** plugin, you can verify your contract's source code and bytecode on the Ethereum blockchain. Here's an example:

```
npx hardhat verify --network sepolia 0xContractAddress
```

By running the preceding command, you can verify the contract with its address on the **Sepolia** network. This helps ensure that the deployed contract matches the source code, enhancing the trustworthiness of your application.

Gas reports and snapshots

Gas reports are an excellent tool for evaluating how smart contracts use computing resources. They also offer cost estimates for each function in your contract and are useful for auditing smart contracts to look for potential optimizations. Forge offers the capability to generate gas reports for your contracts. You have the flexibility to configure which contracts should generate gas reports by utilizing the gas_reports field in the foundry.toml configuration file.

To generate gas reports for specific contracts, you can specify them in the gas_reports field, as in the following:

```
gas_reports = ["MyContract", "MyContractFactory"]
```

If you want to generate gas reports for all contracts, you can use the wildcard character (`*`), as in the following:

```
gas_reports = ["*"]
```

To generate the gas reports, you can execute the `forge test --gas-report` command. This command will initiate the testing process and generate gas reports for the specified contracts.

Additionally, you can combine the gas reporting feature with other subcommands to generate specific gas reports. For example, you can run `forge test --match-test testBurn --gas-report` to generate a gas report specifically for the `testBurn` test case.

By leveraging the gas reporting functionality in Forge, you can gain insights into the gas consumption of your contracts and assess their efficiency. It allows you to focus on specific contracts or tests and make informed decisions to optimize gas usage in your Ethereum projects.

In the Foundry framework, you can utilize Forge to generate gas snapshots for your test functions. Gas snapshots are beneficial for gaining insights into the gas consumption of your contract and comparing gas usage before and after optimizations.

To generate a gas snapshot, simply execute the `forge snapshot` command. By running this command, a file named `.gas-snapshot` will be generated by default. This file contains the gas usage information for all your tests, allowing you to assess the gas consumption of each test function.

Gas snapshots provide a helpful overview of the gas costs associated with your smart contract and aid in identifying areas for potential optimization. By comparing gas usage between different versions or after implementing improvements, you can gauge the impact of your changes on gas efficiency.

A project using Foundry

By now, we have a good understanding of Foundry and all of its core components. We've also talked about most of the important concepts, and it's now time to start applying all that we have learned. We will be doing so by building a small project that will give us hands-on experience of how to build actual projects using Foundry.

In this hands-on project, we will create an NFT, which is a unique digital item like a collectible trading card that is compatible with the popular **OpenSea** platform. We're going to build this with the help of Foundry and also **Solmate**.

> **NFTs**
>
> NFTs are unique digital assets that represent ownership or proof of authenticity of a specific item, artwork, video, music, or other digital content. They are stored on a blockchain, making them secure and easily transferable between individuals.

Solmate is simply a gas-efficient implementation of the ERC721 standard – it is new and modern but is essentially a Solidity smart contract. You can learn more about it at the following link: `https://github.com/transmissions11/solmate/blob/main/src/tokens/ERC721.sol`.

Now that we have an idea of what we're about to build, let's get started.

Getting started

In the *Installing Foundry section*, we installed Foundry using the `foundryup` tool, and that's essentially the first step in this process. The next step is to install the dependencies such as the Solmate implementation for the ERC721 and some **OpenZeppelin** utility libraries.

OpenZeppelin is quite popular in the web3/blockchain space as a collection of industry-standard utility libraries that are open source and free to use. You can read more about it at `https://www.openzeppelin.com/`.

To install the dependencies in your project, all you have to do is run the following command at the root of your project after you use the `forge init` command to start a new project:

```
forge install transmissions11/solmate Openzeppelin/openzeppelin-
contracts
```

The dependencies that are installed are added to the project as Git submodules.

If you followed the instructions correctly, your project should have the `oppenzeppelin-contracts` and `solmate` folders created in the project and you can verify using the `ls` command or the `tree -L 2` command.

A basic NFT

Let's create our NFT contract. In the repository of your forge project created with `forge init`, delete all contracts inside the `src` folder and create a new one called `NFT.sol`. Insert the following code inside it and save:

```solidity
// SPDX-License-Identifier: UNLICENSED
pragma solidity 0.8.10;

import "solmate/tokens/ERC721.sol";
import "openzeppelin-contracts/contracts/utils/Strings.sol";

contract NFT is ERC721 {
    uint256 public currentTokenId;

    constructor(
        string memory _name,
```

```
        string memory _symbol
    ) ERC721(_name, _symbol) {}

    function mintTo(address recipient) public payable returns
(uint256) {
        uint256 newItemId = ++currentTokenId;
        _safeMint(recipient, newItemId);
        return newItemId;
    }

    function tokenURI(uint256 id) public view virtual override returns
(string memory) {
        return Strings.toString(id);
    }
}
```

Let's examine the preceding code. Initially, we import two contracts from our Git submodules. Specifically, we import Solmate's gas-optimized implementation of the ERC721 standard, which our NFT contract will inherit. Within our constructor, we receive the _name and _symbol arguments for our NFT and transmit them to the constructor of the parent ERC721 implementation.

Lastly, we introduce the mintTo function, enabling anyone to create an NFT. This function increments currentTokenId and utilizes the _safeMint function from our parent contract, and the tokenURI function returns the **Uniform Resource Identifier** (**URI**) of a specific token based on its unique identifier (id). This implementation simply converts id (uint256) to a string using the Strings.toString method and returns this string as the token's URI. This function is essential for linking each unique token to its corresponding metadata, which often includes details such as the token's image, name, and attributes.

To compile the NFT contract, first delete the test file of the counter and then execute the forge build command. It is possible to encounter a build failure caused by an incorrect mapping, and the error will look like the following:

```
Error:
Compiler run failed
error[6275]: ParserError: Source "lib/openzeppelin-contracts/
contracts/contracts/utils/Strings.sol" not found: File not found.
Searched the following locations: "/PATH/TO/REPO".
 --> src/NFT.sol:5:1:
  |
5 | import "openzeppelin-contracts/contracts/utils/Strings.sol";
  | ^^^^^^^^^^^^^^^^^^^^^^^^^^^^^^^^^^^^^^^^^^^^^^^^^^^^^^^^^^^^^
  |
```

To resolve this issue, you can rectify it by configuring the appropriate remapping. Simply create a file called `remappings.txt` within your project and include the following line:

```
openzeppelin-contracts/=lib/openzeppelin-contracts/
```

> **Incompatible Solidity versions**
>
> You may also receive an error that states that the Solidity versions are incompatible: "**Found incompatible Solidity versions...**" In that case, you should change the Solidity version under `pragma solidity 0.8.13` to the recommended one.

By default, the compiler output will be stored in the `out` directory. To deploy the compiled contract using Forge, you need to set environment variables for the **RPC** endpoint and the **private key** (**PK**) you wish to use for deployment.

An RPC endpoint is a crucial component in blockchain technology, acting as a gateway for interacting with a blockchain network. It allows your application to communicate with the blockchain, sending transactions, querying balances, and performing other network interactions. Without a properly configured RPC endpoint, your application would not be able to deploy or interact with contracts on the blockchain network. In summary, the RPC endpoint is the link between your deployment tool (such as Forge) and the blockchain itself, enabling the deployment and management of smart contracts.

To make the RPC endpoint work for deploying a smart contract using Forge, you typically need to set up an account with a blockchain infrastructure provider, such as Infura or Alchemy. These providers offer access to blockchain networks through their **Remote Procedure Call** (**RPC**) endpoints. Once you have an account, you can create a project on their platform and obtain an RPC URL specific to your project. This URL is then used as the RPC endpoint in your environment variables. By setting this along with your PK as environment variables, your deployment tool executes the following commands:

```
export RPC_URL=<Your RPC endpoint>
export PRIVATE_KEY=<Your wallet's private key>
```

It's important to handle the PK securely. Typing your PK directly into the terminal is not recommended, as it can leave traces in your `bash`/`zsh` history. A safer approach is to store the PK in a script file, ensuring that the file is included in `.gitignore` to prevent it from being accidentally committed to version control. This way, you can run the script when needed without having to type your PK directly, enhancing security. For this test project, while it might be acceptable to use a PK in this manner, it's crucial to remember that this is not a good practice for production environments.

Once the environment variables are set, you can deploy your NFT with Forge by running the following command, ensuring you include the relevant constructor arguments for the NFT contract:

```
forge create NFT --rpc-url=$RPC_URL --private-key=$PRIVATE_KEY
--constructor-args <name> <symbol>
```

If the deployment is successful, you will see the deploying wallet's address, the contract's address, and the transaction hash displayed in your terminal.

Testing the program

Now that we've built our program, let's test it out by interacting with our smart contract.

Interacting with functions on your NFT contract is made effortless with Cast, which is Foundry's command-line tool designed for smart contract interaction, transaction sending, and obtaining chain data. Let's explore how we can utilize Cast to mint NFTs from our NFT contract.

Assuming you have already set your RPC and PK environment variables during the deployment process, you can mint an NFT from your contract by executing the following command, replacing `<arg>` with your address:

```
cast send --rpc-url=$RPC_URL <contractAddress> "mintTo(address)" <arg>
--private-key=$PRIVATE_KEY
```

Congratulations! You have successfully minted your first NFT from your contract. To verify the ownership of the NFT with a `currentTokenId` value of `1`, you can perform a sanity check by running the following Cast call command. The address you provided earlier should be returned as the owner:

```
cast call --rpc-url=$RPC_URL --private-key=$PRIVATE_KEY
<contractAddress> "ownerOf(uint256)" 1
```

We have now built and interacted with our project. But that's not all; we can also get the gas reports from our project, and that's exactly what we will do in the next section.

Gas reports

Foundry offers comprehensive **gas reports** for your contracts, providing details about the gas costs associated with each function called during testing. The report includes information such as the minimum, average, median, and maximum gas costs. To generate the gas report, simply execute the following command:

```
forge test --gas-report
```

This feature proves beneficial when analyzing various gas optimizations within your contracts.

In the context of optimizing gas efficiency in our ERC721 implementation, we have conducted a comparative analysis using reports obtained by running `forge test --gas-report` on that repository between the utilization of OpenZeppelin and Solmate libraries. The following screenshots offer a comprehensive view of the gas consumption achieved through these implementations. The first screenshot highlights the gas utilization of the Solmate implementation.

NftSolmate contract					
Deployment cost	**Deployment size**				
1583919	8489				
Function name	**min**	**avg**	**median**	**max**	**# calls**
balanceOf	635	635	635	635	2
currentTokenId	2308	2308	2308	2308	1
mintTo	473	44574	69399	72745	9
ownerOf	566	566	566	566	1

Figure 6.1 – With Solmate's ERC721 implementation

The second screenshot demonstrates the gas consumption of the OpenZeppelin implementation. You can find the NFT implementation utilizing both libraries in this repository.

NftOZ contract					
Deployment cost	**Deployment size**				
1791936	9528				
Function name	**min**	**avg**	**median**	**max**	**# calls**
balanceOf	705	705	705	705	2
currentTokenId	2308	2308	2308	2308	1
mintTo	473	44797	69632	73214	9
ownerOf	602	602	602	602	1

Figure 6.2 – With OpenZeppelin's ERC721 implementation

As evident from the report in *Figure 6.2,* our implementation using Solmate exhibits a reduction of approximately 500 gas in successful mint operations (representing the maximum gas cost for the `mintTo` function calls).

There you have it! I trust that this will provide you with a solid foundation for getting started with Foundry. I firmly believe that there's no better approach to truly comprehending Solidity than by writing your tests in Solidity itself. This approach also minimizes the need for frequent context switching between JavaScript and Solidity.

Summary

In this chapter, we delved into the Foundry framework, a powerful toolset for Ethereum development. The chapter commenced by introducing Forge, an essential component of Foundry that simplifies the compilation and deployment of smart contracts. We learned how to use Forge to compile contracts and handle deployment configurations effortlessly. The flexibility offered by Forge allows for seamless integration with different networks and deployment environments.

Next, we explored Cast, Foundry's command-line tool designed for interacting with smart contracts. We discovered how Cast simplifies the process of sending transactions, querying contract data, and interacting with functions. The ability to specify RPC endpoints and PKs enables smooth contract interaction from the command line.

Anvil, another vital component, showcased the gas reporting capabilities of Foundry. We learned how Anvil generates comprehensive gas reports for contracts, providing valuable insights into gas costs for individual functions. This information proved invaluable for optimizing contract performance and identifying areas for improvement.

We also examined Chisel, a testing framework that facilitates writing and executing unit tests for smart contracts. We discovered the advantages of writing tests in Solidity itself, reducing the need for context switching between JavaScript and Solidity. Chisel's integration with other Foundry components ensures a seamless testing experience.

Finally, we built our own project using all of the concepts we have learned and we saw how to build, deploy, test, verify, and get gas reports using Foundry, and all these skills are extremely important for working with blockchains.

Overall, this chapter equipped us with a thorough understanding of the Foundry framework and its core components. By leveraging Forge, Cast, Anvil, and Chisel, developers can enhance their Ethereum development workflow, increase efficiency, and build robust smart contracts with ease.

In the next chapter, we will learn about a popular blockchain that uses Rust – Solana. We will also build a small dApp using Solana, so stay tuned.

7

Exploring Solana by Building a dApp

In this chapter, we'll delve into the intricate world of Solana's development landscape, focusing on foundational aspects crucial for building robust programs on blockchain. With a focus on practical implementation and meticulous validation, we aim to provide a comprehensive understanding of essential concepts, empowering developers to build resilient and efficient solutions on the Solana network. Before diving into the practical aspects of building a **decentralized application (dApp)**, it's crucial to comprehend the fundamental concept of a dApp. Hence, in this chapter, let us lay the groundwork by understanding precisely what constitutes a dApp.

Exploring nuances of account structuring, data sizing, and validation mechanisms, we'll journey through the following topics:

- Introducing dApps
- Setting up the environment for Solana
- Working with Solana frameworks and tools
- Building and deploying a dApp
- Creating accounts for our custom dApp
- Creating our first instruction
- Implementing logic
- Creating tests for our instructions

Introducing dApps

dApps are digital tools and platforms that operate without a central authority, much like the apps we use every day, but with a twist—they are built on the backbone of blockchain technology. Think of them as software applications that do not rely on a single company or server to function. Instead, they run on decentralized networks where data is securely stored and transactions are transparently recorded across multiple computers, ensuring security and reliability without the need for intermediaries. These dApps open doors to a world where users can interact directly with services, trust is built on transparency, and decision-making often lies in the hands of a diverse community. Let us move on and understand what dApps are in more depth.

What are dApps?

dApps represent a major shift in how software applications are built and function. Unlike traditional apps, which rely on centralized servers, dApps use blockchain technology as their core foundation. This approach allows them to operate on decentralized networks, where no single entity has total control:

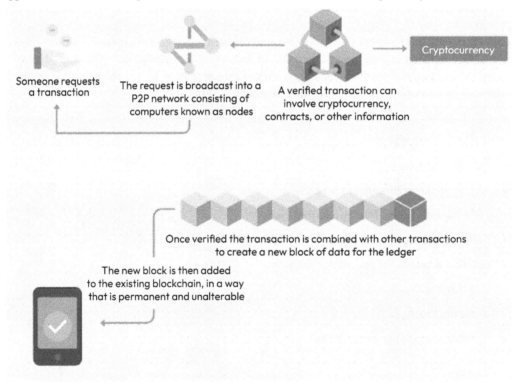

Figure 7.1 – Expanded concept of dApps

Essentially, dApps are changing the way we think about software. Take a look at *Figure 7.1*, and you can see dApps run on a network of computers rather than just one server, making use of blockchain's secure and distributed ledger. This setup means that dApps can function independently without being controlled by any one organization or person. This new way of building applications is not just innovative but also offers a more autonomous and democratic digital environment. Instead, they thrive on consensus mechanisms intrinsic to blockchain networks, enabling a democratic and transparent operational environment.

Types of dApps

Let's take a look at the different types of dApps available:

- **Financial dApps (DeFi)**: Financial dApps, commonly known as **decentralized finance** (**DeFi**) applications, form a cornerstone within the dApp ecosystem. These pioneering applications center on offering an array of financial services, leveraging the transformative potential of blockchain technology. Examples include the following:

  - **Decentralized exchanges (DEXs)**: Platforms facilitating **peer-to-peer** (**P2P**) trading of digital assets without the need for intermediaries

  - **Lending protocols**: Systems enabling individuals to lend or borrow digital assets directly from others without the involvement of traditional financial intermediaries

  - **Yield farming platforms**: Mechanisms that allow users to earn rewards or yields by providing liquidity to DeFi protocols

  - **Stablecoins**: Cryptocurrencies pegged to stable assets such as fiat currencies (analogous to USD), providing stability amid the volatility of the cryptocurrency market

- **Utility dApps**: Utility dApps extend beyond the realm of financial services, offering an array of utilities that transcend monetary applications. These versatile applications encompass various functionalities, including the following:

  - **Decentralized storage**: Platforms that leverage blockchain for secure and decentralized storage solutions, ensuring data integrity and accessibility.

  - **Identity verification**: Applications using blockchain for secure and tamper-proof identity verification and management systems.

  - **Prediction markets**: Prediction markets are platforms that allow users to speculate on the outcomes of real-world events, leveraging decentralized consensus mechanisms to ensure fairness and transparency. Beyond prediction markets, oracles play a crucial role in the broader ecosystem of dAppS by serving as bridges between blockchain networks and external data sources. **Oracles**, such as **Chainlink**, provide vital, real-world information that smart contracts can rely on, making them an essential component of many blockchain-based applications, potentially even more critical than prediction markets themselves.

- **Decentralized governance systems**: Applications that enable decentralized decision-making and governance processes, often through voting mechanisms

- **Gaming and entertainment dApps**: The realm of gaming and entertainment has witnessed a significant evolution with the emergence of dApps. These applications leverage blockchain technology to offer the following:

  - **Games and collectibles**: Innovative games leveraging blockchain for asset ownership, provable scarcity, and interoperability between different gaming ecosystems

  - **Entertainment platforms**: Platforms offering diverse entertainment content, leveraging blockchain for provenance tracking, copyright protection, and decentralized distribution

As we delve deeper into the spectrum of dApps, it becomes evident that their diversity extends far beyond financial services, encompassing utilities, entertainment, and governance. These varied categories underscore the versatility and transformative potential of dApps, catering to a wide array of user needs within the decentralized ecosystem.

Benefits of dApps

The adoption and proliferation of dApps bring forth a multitude of advantages:

- **Decentralization and immutability**: dApps thrive on decentralized networks, eradicating vulnerabilities stemming from **single points of failure** (**SPOFs**). By storing data across a distributed ledger, dApps ensure robust data integrity and reliability. This decentralized architecture enhances security, eliminating risks associated with centralized control and fostering a resilient ecosystem.

- **Enhanced transparency**: At the heart of dApps lies the transparent nature of blockchain technology. All transactions and operations within these applications are visible and auditable by all participants. This transparency cultivates an environment of trust and accountability, mitigating the need for blind trust in intermediaries.

- **Elimination of intermediaries**: By circumventing intermediaries, dApps streamline processes, significantly reducing associated costs and enhancing operational efficiency across diverse industries. Sectors such as finance, supply chain management, and legal domains witness heightened efficiency and reduced friction through the elimination of intermediaries.

- **Community governance**: A distinguishing feature of many dApps is their facilitation of decentralized governance models. These models empower community members to actively participate in decision-making processes, fostering a democratic and inclusive approach to managing applications. This democratized governance framework ensures broader consensus and enhances user engagement.

- **Global accessibility**: Running on decentralized networks, dApps transcend geographical barriers, ensuring global accessibility. Individuals worldwide can interact with these applications without constraints, democratizing access to services and functionalities on a global scale.

Now that we have understood what a dApp is, let's embark on an exciting journey into the world of Solana-Anchor to create our own dApps! Our first step begins with setting up the environment essential for crafting innovative dApps on the Solana blockchain. To kick things off, we'll dive into the next section. Here, we will walk through the meticulous process of installing and configuring the fundamental tools necessary for a conducive development environment.

Setting up the environment for Solana

Before diving into the intricacies of Solana's development ecosystem, setting up the right environment lays the foundation for seamless programming experiences. In this section, we'll navigate through the setup process, ensuring all necessary tools, libraries, and configurations are in place. From installing essential software to configuring development environments, this initial step is pivotal in facilitating smooth and efficient Solana development. Let's embark on this journey to configure the ideal environment for building robust and scalable blockchain solutions.

Installing Rust

Rust stands as a versatile systems programming language renowned for its emphasis on safety, speed, and concurrency. It plays a pivotal role in building Solana programs, including smart contracts and dApps. The utilization of Rust within the Solana ecosystem ensures robustness and security without compromising performance. Rust's memory safety guarantees, devoid of a **garbage collector** (**GC**), fortify the integrity of Solana's code base, making it a preferred language for developing complex and secure applications.

Installation steps

Here are the steps listed for Rust installation on Unix, macOS, and Windows:

1. **Install rustup**:

   - **Unix or macOS**: To set up Rust, including `rustc` (the Rust compiler) and `cargo` (Rust's package manager and build tool), run the following command in your terminal and follow the onscreen instructions to complete the installation process:

     ```
     curl --proto '=https' --tlsv1.2 -sSf https://sh.rustup.rs |
     sh
     ```

   - **Windows**: The installation process involves downloading and running the `rustup-init.exe` file from the official Rust website. Please follow these steps:

     i.   Download `rustup-init.exe` from `https://rustup.rs/`.

     ii. Run the downloaded file and follow the onscreen instructions to install Rust. This includes choosing installation options and setting your PATH variable.

     iii. After installation, you might need to restart your terminal or Command Prompt to ensure the `rustc`, `cargo`, and `rustup` commands are available in your PATH variable.

2. **PATH variable added to shell config**: After installation, it incorporates the subsequent PATH variable — or a similar equivalent — into your shell configurations. This detail proves useful should you decide to relocate it to your dotfiles or a similar setup:

```
export PATH="$HOME/.cargo/bin:$PATH
```

Windows users will typically find that rustup automatically configures the PATH variable during installation.

3. **Check the Rust version**: To verify successful installation and display the Rust compiler version, use the following command:

```
rustup --version
rustc --version
```

4. Additionally, confirm the installation of cargo, the Rust package manager, by entering the following:

```
cargo --version
```

The installation of Rust encompasses rustup for managing Rust versions, rustc for compiling Rust programs, and cargo for managing project dependencies and building projects. These components collectively provide a robust foundation for Solana program development in Rust. For further details on installation on Unix, macOS, and Windows, you can follow the directions at https://forge. rust-lang.org/infra/other-installation-methods.html.

Introducing Solana

Solana is a leading blockchain platform known for its high speed and ability to handle a large number of transactions. It tackles common problems in traditional blockchains, such as slow speeds and limited scalability. Solana combines new technologies and innovative methods to process transactions quickly and support smart contracts with minimal delay.

In a field where slow transactions are a common issue, Solana stands out by offering a solution to these scalability challenges. It uses advanced technology and new protocols to allow more transactions without slowing down. This approach marks a significant change in blockchain design, achieving both speed and scalability without compromising security or decentralization. Solana's innovative architecture opens up new possibilities for dApps and broadens their potential uses, setting a new standard in blockchain technology.

The network's bedrock innovation, the **Proof of History (PoH) consensus mechanism**, operates as a timestamping method that sequences transactions, optimizing throughput without compromising accuracy or security. This ingenious approach to establishing chronological order without the computational intensity of traditional consensus algorithms is the linchpin of Solana's exceptional scalability.

Inherent challenges of scalability that have long plagued blockchain systems find resolution within Solana's architectural marvel. Transactions traverse the network seamlessly, achieving speeds that had previously been elusive in the blockchain realm. The network prides itself on the ability to process thousands of **transactions per second** (**TPS**) at a fraction of the cost compared to legacy systems.

Understanding Solana's innovative architecture

Solana's architecture is a testament to its ambition to revolutionize blockchain technology. At the heart of its efficiency lies the groundbreaking PoH mechanism. PoH introduces a verifiable time function that provides a chronological order for transactions without necessitating the computational workload of traditional consensus mechanisms. This innovation significantly reduces the time required to validate transactions, enhancing the network's scalability without compromising decentralization:

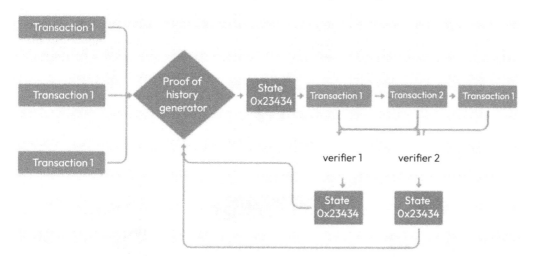

Figure 7.2 – Solana's network architecture

Figure 7.2 shows how transactions work in the Solana blockchain. Transactions start at the left, where they are put into the network. These transactions are then processed by Solana's PoH generator, a key component of its high-speed blockchain. The PoH generator sequences transactions, creating a historical record that proves the time and order of each transaction without needing the agreement of all nodes in the network. The state, represented by the 0x23434 hex code, likely signifies the current state of the blockchain ledger after transactions have been processed by the PoH generator. This state is then subject to verification by nodes within the network, labeled here as **verifier 1** and **verifier 2**. These **verifiers** check the transactions against the blockchain's current state to ensure they are valid and then apply them to their local copies of the ledger, updating the state to reflect the changes made by the transactions.

This system allows Solana to process and confirm transactions rapidly, providing an architecture where throughput scales with the number of verifiers, thereby addressing scalability issues faced by earlier blockchains. The PoH component serves as a cryptographic timestamp, ensuring the sequence of transactions is recorded and verifiable, which contributes to the overall security and efficiency of the network.

In summary, the architecture depicted in the diagram showcases how Solana leverages PoH for quick transaction processing and verification, maintaining an accurate and secure ledger state across its decentralized network. This structure underpins Solana's ability to offer fast and scalable blockchain solutions for a wide array of dApps.

Why Solana?

Solana's magnetism stems from its capacity to furnish a resilient infrastructure that gracefully navigates the **scalability trilemma** - an intricate balance between scalability, decentralization, and security that most blockchain networks struggle to achieve simultaneously. Solana achieves its impressive balance of speed, scalability, and security through the integration of several innovative technologies:

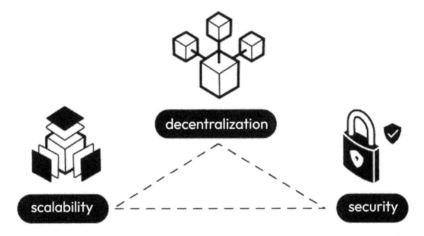

Figure 7.3 – Balance between scalability, decentralization, and security

PoH is a key feature that allows the network to keep track of the order of transactions, serving as a cryptographic timestamp that streamlines the process of verifying transactions. This means that validators on the network can process transactions without waiting for consensus on transaction order, significantly speeding up the confirmation time.

The **Tower consensus protocol** builds upon PoH, providing additional security by ensuring validators agree on the same order of transactions and preventing them from voting on multiple conflicting versions of the blockchain. **Gulf Stream** pushes transactions to validators even before the previous batch of transactions is finalized, which reduces confirmation time and the memory pool size of unconfirmed transactions on validators. This protocol allows Solana to support a high number of TPS without getting bogged down. Lastly, **Turbine** is a block propagation protocol that breaks data into smaller packets, making it easier to transmit and manage across the network. This makes the network more efficient, especially when handling large volumes of transactions.

Together, these technologies allow Solana to process transactions rapidly and at scale while ensuring the network remains secure and resilient. By cleverly combining these protocols, Solana can handle the high demands of modern dApps without the usual trade-offs that limit other blockchains.

Key features of Solana

Let's now look at some key features of Solana:

- **Speed and scalability**: The cornerstone of Solana's allure rests upon its unparalleled ability to process transactions at a breathtaking speed, soaring to an impressive 65,000 TPS. This monumental throughput, coupled with nominal fees, lays the foundation for a network that doesn't compromise on efficiency while catering to a myriad of dApps.

- **PoH**: Solana's ingenious PoH mechanism stands as a testament to its innovative architecture. By timestamping transactions, PoH fosters network scalability without sacrificing decentralization. This novel approach obviates the need for arduous consensus protocols, augmenting Solana's ability to process a colossal number of transactions without compromising on precision.

- **Tower consensus**: Solana intertwines **Proof of Stake (PoS)** consensus with PoH in the Tower consensus mechanism. This amalgamation ensures a secure and efficient environment for block production, fortified by the staking of **SOL (Solana) tokens**. This synergy fortifies the network's resilience, underscoring Solana's commitment to a secure and decentralized ecosystem.

- **Gulf Stream**: Gulf Stream represents a pivotal mechanism for optimizing network bandwidth by adeptly propagating blocks across the network. This strategic optimization further amplifies Solana's performance, reducing bottlenecks and bolstering overall network efficiency.

- **Turbine**: Turbine serves as a block propagation protocol meticulously designed to streamline the dissemination of blocks throughout the network. This protocol significantly reduces latency, enhancing overall efficiency and fortifying Solana's position as an agile and responsive blockchain platform.

Solana's robust feature set and meticulously engineered architecture position it as an irresistible choice for developers seeking to build dApps that demand top-tier throughput and ultra-low latency. Its innovative blend of technologies not only addresses the trilemma but transcends it, opening doors to a new era of scalable and high-performance blockchain networks.

Installing Solana

The Solana installation process is designed to equip developers with the necessary command-line tools and a local network environment for developing and testing dApps efficiently. Here are instructions for how to install it on different platforms:

- **For macOS users**: To install Solana on macOS, using Homebrew is recommended for its simplicity and for ensuring you receive the latest version of Solana. If you encounter an error related to `gnu-tar`, follow the next steps to resolve it:

  I.   Install Solana using Homebrew by running the following command in your terminal:

  ```
  brew install solana
  ```

  II. If you encounter a `gnu-tar` error, install `gnu-tar` separately using Homebrew:

  ```
  brew install gnu-tar
  ```

  III. After installing `gnu-tar`, add it to your `PATH` variable by adding the following line to `~/.zshrc` or `~/.bash_profile` (depending on the shell you use):

  ```
  export PATH="/usr/local/opt/gnu-tar/libexec/gnubin:$PATH"
  ```

  IV. Restart your terminal to apply the changes.

- **For Linux users**: Linux users should download and install the pre-built versions of Solana available on the official Solana GitHub releases page, following the provided instructions for their specific distribution.

- **For Windows users**: Windows users are directed to follow the detailed instructions provided on the official Solana documentation website to ensure a correct and efficient installation process. Please visit `https://docs.solana.com/cli/install#windows` for a step-by-step guide.

- **Post-installation steps**: After installation, Solana will be integrated into your shell's `PATH` variable automatically, simplifying access to its functionalities. If required, manual `PATH` variable updates will be instructed by the installer:

  ```
  export PATH="$HOME/.local/share/solana/install/active_release/
  bin:$PATH"
  ```

- **Validating your installation**: To confirm a successful installation, check the Solana version:

  ```
  solana --version
  ```

- Additionally, you can initiate a local validator to verify the setup (use *Ctrl + C* to exit), which will create a `test-ledger` directory in your current directory:

   ```
   solana-test-validator
   ```

- This confirms that Solana is ready for local development activities.

We will now see how we can use Solana locally.

Using Solana locally

Utilizing Solana in a local environment entails the initiation of a dedicated Solana network within your development setup. This practice holds paramount importance as it serves as a controlled space for testing and deploying applications, ensuring a secure and controlled atmosphere conducive to development.

The rationale behind deploying Solana locally rests upon its ability to emulate the blockchain environment without necessitating interaction with `mainnet`. This emulation provides developers with a vital sandboxed space for rigorously testing and meticulously debugging applications before their deployment into live environments.

To initiate a local Solana network, the straightforward command `solana-test-validator` is employed for macOS 13 and above; you need to install `gnu-tar` for this to work. This action initializes the local Solana network and concurrently generates a dedicated directory labeled `test-ledger` within your present working directory.

Once the local Solana network is operational, verifying its status is achievable through the `solana cluster-version` command. This step is instrumental in confirming the successful initiation and functionality of the local Solana network, ensuring its readiness for subsequent development and testing tasks.

By utilizing Solana locally and executing these critical steps, developers gain a controlled environment for meticulous testing and debugging, fostering a robust application development cycle before deploying applications to production or live networks.

Let us now generate our local key pair.

Generating a local key pair

The process of generating a **local key pair** involves the creation of cryptographic keys imperative for seamless interactions within the Solana network. These keys play a pivotal role in facilitating secure transactions and authorizing various actions on the blockchain.

The generated key pair encompasses both a public and private key, serving as the cornerstone of security in Solana transactions. The **private key** allows for the authentication of transactions, while the **public key** enables verification and serves as the address for receiving assets or interacting with the network.

Generation steps

Before proceeding, it's prudent to verify whether you already possess a local key pair. You can do so by executing the following command:

```
solana address
```

To generate a new key pair, initiating the cryptographic key creation process involves executing the subsequent command:

```
solana-keygen new
```

Following the generation, to view your public key from the generated key pair, utilize the next command. Replace `path/to/keypair.json` with the appropriate path to your key-pair file:

```
solana-keygen pubkey path/to/keypair.json
```

Ensuring the presence and accessibility of a local key pair is paramount for engaging in secure and authorized interactions within the Solana blockchain. This foundational step establishes the cryptographic foundation necessary for transactions and engagement within the Solana network.

Moving on, in the next section, we will look at the setup of essential frameworks and tools pivotal for Solana development.

Working with Solana frameworks and tools

In this section, we'll focus on establishing essential frameworks and tools critical for Solana development. This includes installing and configuring specialized frameworks such as Anchor, a key tool that simplifies smart contract development on Solana. We'll delve into setting up necessary dependencies, exploring the functionalities of these frameworks, and configuring them to optimize our development workflow. By ensuring the right frameworks and tools are in place, we lay the groundwork for efficient and effective Solana blockchain development. Let's embark on configuring these vital elements for a streamlined development journey, starting with Anchor.

Introducing Anchor

Anchor, positioned as a sophisticated and versatile framework, plays a pivotal role in the realm of dApp development within the Solana blockchain ecosystem. Functioning as an essential bridge between developers and the multifaceted features of Solana, Anchor stands distinguished by providing an expansive suite of tools and libraries meticulously crafted to expedite and enrich the development life cycle.

Anchor is a development framework that simplifies blockchain programming. Its main goal is to take away the complex details of blockchain technology so that developers can focus on creating new and innovative applications. Anchor is especially helpful for those new to blockchain development, as well as experienced developers looking for a more straightforward way to build dApps.

The framework is user-friendly and well documented, making it accessible to developers of various skill levels. With its comprehensive tools and resources, Anchor is an ideal starting point for those new to blockchain and a helpful tool for experienced developers to enhance their projects.

In essence, Anchor makes developing on Solana's powerful blockchain easier. It provides clear guidance and support, helping developers unlock the full potential of blockchain technology without getting lost in its complexity

Let us now look at some of the benefits of Anchor:

- **Simplified development**: Anchor provides high-level **application programming interfaces (APIs)** and tools that simplify Solana's complex architecture, allowing developers to focus on creating innovative dApps more efficiently.

- **Efficiency and speed**: Anchor enhances dApp development on Solana by abstracting complexities and reducing boilerplate code, which accelerates development cycles. While Solana's speed benefits all dApps, Anchor specifically speeds up the development process through its efficient framework, enabling quicker project completion.

- **Enhanced security and reliability**: With Anchor, developers get an extra layer of security by abstracting critical operations, leading to safer and more reliable dApps.

- **Comprehensive documentation and community support**: Anchor offers extensive resources and a supportive community for developers to learn, solve problems, and innovate together.

Apart from several advantages Anchor, Anchor functions with some key functionality, which we discuss next.

Key concepts in Anchor

Let us look at some important concepts in Anchor:

- **Programs and transactions**: Anchor revolves around the creation of programs that interact seamlessly with Solana's blockchain via transactions. This core principle simplifies the intricate process of crafting and executing transactions, allowing developers to define program logic with utmost efficiency. By abstracting the complexities underlying transaction execution, Anchor empowers developers to focus on implementing and refining programmatic functionalities.

- **Interface Description Language (IDL)**: The employment of IDL within Anchor plays a pivotal role in defining on-chain program instructions and corresponding data structures. This robust feature serves as a bridge, streamlining interaction between on-chain programs and off-chain applications. IDL ensures a standardized method of communication, facilitating seamless integration and interoperability between different layers of dApp architecture.

- **Context and state**: Central to Anchor's functionality is its adept management of program context and state. This capability enables fluid interactions between on-chain and off-chain components of dApps. By abstracting the complexities inherent in state management, Anchor ensures a level of consistency and reliability critical to the seamless functioning of dApps, thus bolstering their robustness and usability.

- **Client library and CLI**: Anchor's provision of a comprehensive client library and **command-line interface** (**CLI**) stands as a testament to its commitment to empowering developers. These tools simplify the deployment, testing, and management of Solana-based dApps, offering developers an intuitive and efficient workflow. The client library and CLI encapsulate essential functionalities, providing developers with a cohesive and streamlined environment for building and deploying applications.

Anchor simplifies Solana dApp development by providing developers with easy-to-use tools, comprehensive functionalities, and thorough documentation. It stands out in the Solana ecosystem for its ability to demystify blockchain complexities and boost development efficiency. As a result, Anchor is not just a facilitator for building applications; it actively encourages innovation, helping developers create effective dApps on the Solana blockchain with less effort. Its support system and resources make it a go-to framework for both new and experienced developers working with Solana.

Installing Anchor

To install the Anchor CLI, commence by executing the following command:

```
cargo install --git https://github.com/project-serum/anchor anchor-cli
--locked
```

This command orchestrates the installation of the Anchor CLI, facilitating its accessibility within your development environment.

Once the installation concludes, confirming the installed version of Anchor is achievable through the following command:

```
anchor --version
```

The installation of Anchor propels developers into a domain where complexities of smart contract development are abstracted, fostering a conducive environment for streamlined and efficient creation and deployment of Solana-based dApps.

Let us now install Yarn, which will conclude the prerequisites.

Installing Yarn

Yarn operates as a robust package manager primarily utilized for handling dependencies within JavaScript projects, particularly in the context of React applications that interface with Solana-based dApps.

The utility of Yarn lies in its efficiency and effectiveness in managing JavaScript packages and dependencies. It simplifies the installation, management, and maintenance of various packages, ensuring a seamless and hassle-free development experience within Solana-based projects.

Yarn installation is versatile, allowing multiple approaches across different operating systems.

For npm-based installation, run the following command:

```
npm install -g yarn
```

On Mac systems utilizing Homebrew, run this command:

```
brew install yarn
```

For Linux systems using apt, run the following command:

```
apt install yarn
```

Following the installation, confirming the version of Yarn installed can be accomplished through this command:

```
yarn --version
```

Each aforementioned installation method plays a pivotal role in configuring the development environment, enabling a smooth and efficient setup for developers venturing into Solana dApp development using Anchor. These detailed instructions aim to streamline the setup process, providing readers with a robust foundation for their Solana-based projects.

With the completion of the installation of the required frameworks and tools, let us now move on and create a project for our dApp.

Creating a new Anchor project

The installation and configuration of essential frameworks and tools lay the groundwork for seamless Solana dApp development. Now that we've completed the setup, let's dive into creating our dApp project using Anchor.

To start our dApp journey with Anchor, let's initiate a new project named solana-custom. First, navigate to your designated development folder, commonly located in ~/Code or in your preferred directory:

```
# Navigate to your development folder.
cd ~/Code
# Create a new Anchor project named "solana-custom."
anchor init solana-custom
# Move into the newly created project.
cd solana-custom
```

Upon executing these commands, Anchor crafts the foundational structure for our project:

- **Programs folder**: This directory houses Solana programs. It comes preloaded with a basic program, reducing the need for extensive scaffolding.

- **Tests folder**: Designed for JavaScript tests directly interacting with our programs. It includes a pre-generated test file for the auto-created program.

- **Anchor.toml configuration file**: Essential for configuring program IDs, Solana clusters, and other project settings.

- **Empty app folder**: Although currently empty, this directory will eventually house our JavaScript client code.

Having scaffolded our project, we are now prepared to explore the developmental cycle of building a Solana program. In the subsequent sections, we will delve into deploying, testing, and understanding the default program generated by Anchor. This comprehensive exploration will provide invaluable insights into the intricate aspects of developing a Solana-based application.

Through the well-organized structural setup, Anchor substantially streamlines the initial stages of Solana dApp development. Its preconfigured elements empower developers to concentrate on core functionalities and innovative aspects of their dApps.

Building and deploying a dApp

This segment serves as a comprehensive guidebook to the real-world application of Solana dApps. It is a journey through the nitty-gritty of crafting, refining, and managing dApps leveraging Anchor's CLI. The section outlines indispensable procedures fundamental for initiating, deploying, and refining dApps, all while employing the versatile toolkit that the Anchor CLI offers. We will delve into intricate nuances of building and deploying dApps, establishing and navigating a local Solana ledger setup to rigorously test and validate, administering and revising program IDs for seamless and efficient functionality, and harnessing capabilities of Anchor scripts to elevate dApp functionalities and automate critical tasks. Each step is meticulously crafted to impart not just theoretical insight but hands-on practical experience, empowering developers with the expertise required to fashion robust and dependable Solana-based dApps.

Building and deploying with Anchor

Anchor simplifies the intricacies of deploying Solana programs through its two pivotal commands - `anchor build` and `anchor deploy`. These commands act as the backbone of the development process, facilitating the compilation and deployment of programs onto the Solana network seamlessly.

Compiling using anchor build

Executing `anchor build` initiates the compilation process for the program. This step involves the Rust compiler's meticulous examination of the code, which may result in warnings or errors. For instance, you might encounter messages such as `unused variable: 'ctx'`. Typically, these warnings stem from the simplicity of the autogenerated program, where certain variables might not be actively utilized. However, they pose no significant concerns. Upon successful compilation, the `target` folder gets updated, to house built releases and deployment artifacts. This `target` folder serves as a local repository exclusive to your machine and remains outside the **version control system** (**VCS**):

```
# Compile your program.
anchor build
```

Moreover, `anchor build` generates an IDL file—a JSON document containing comprehensive specifications of your Solana program. This IDL file encapsulates essential information, including instructions, parameters, and generated accounts, ensuring structured interactions between your Solana program and its JavaScript client.

Deploying using anchor deploy

The `anchor deploy` command dispatches the latest build to the Solana cluster for deployment. When a program is initially built, it generates a public and private key pair stored in the `target` directory. This public key becomes the unique identifier or program ID for your Solana program:

```
# Deploy your compiled program.
anchor deploy
```

However, attempting deployment without a preconfigured network leads to an error—`error sending request for url (http://localhost:8899/)`. To rectify this, it's imperative to set up a local ledger environment.

Running a local ledger

A **local Solana ledger** functions as a simulated Solana cluster within your local environment, facilitating development without interacting with the live Solana blockchain. This emulation is instrumental for testing and validation purposes, enabling developers to deploy and assess their applications locally.

Initiating a local ledger is conveniently accomplished by executing the following command in your terminal:

```
solana-test-validator
```

This command keeps an active session in your terminal until you exit using *Ctrl + C*. With the session active, you have a local ledger ready for deployment. Before deploying your applications onto this local ledger environment with `anchor deploy`, you must ensure your wallet has the necessary funds. This is achieved by airdropping SOL tokens to your wallet using the Solana CLI:

1. To point the Solana CLI to your local test validator, run the following command:

    ```
    solana config set --url localhost
    ```

2. After setting the URL, airdrop SOL to your wallet by executing this command:

    ```
    solana airdrop
    ```

    This command airdrops one SOL to your wallet, which is the required balance for executing transactions on the local network. Note that you can airdrop one SOL at a time.

With your wallet funded, you're now ready to deploy your applications using `anchor deploy`, interacting with your local ledger environment effectively.

However, it's crucial to note that all data sent to your local ledger gets stored in a `test-ledger` folder created within the current directory. To prevent this folder from being committed to your Git repository, update your `.gitignore` file as follows:

```
.anchor
.DS_Store
target
**/*.rs.bk
node_modules
test-ledger
```

Exiting the local ledger using *Ctrl + C* won't erase any data sent to the cluster. However, deleting the `test-ledger` folder will result in data removal. Alternatively, you can achieve the same outcome by employing the `--reset` flag:

```
solana-test-validator --reset
```

> **Note**
>
> To reset the state of the Solana test validator, ensuring a fresh start for development testing, the `--reset` flag can be used.
>
> Alternatively, for convenience, the shorter `-r` flag can also be used.

This emulation environment serves as a valuable asset for testing Solana-based applications without affecting the live blockchain. By seamlessly integrating this simulated cluster, developers gain a controlled space for robust testing and validation.

Updating the program ID

After deploying your program initially using `anchor build` and `anchor deploy`, it becomes crucial to update the program ID to ensure a consistent and accurate identifier across various configurations and files within your Solana project.

During the initial deployment, a new key pair is generated, containing a public address necessary for your program's identification. However, this public address is not readily known right after deployment. You can obtain it using the following Solana command:

```
solana address -k target/deploy/solana_custom-keypair.json
# Example output: 6xyMvrMKdjbXuTSjaDULm2vit8Uq6MjCgJj5N6zpJASY
```

Please note that the `solana_custom-keypair.json` filename might vary based on the naming conventions set within your project.

Once you have retrieved the program ID, it becomes essential to update it in specific locations within your Solana project. Anchor uses these locations as initial placeholders during the project's setup. Here are the steps to update these locations:

1. **Anchor.toml configuration file**: Navigate to the `Anchor.toml` configuration file and locate the `[programs.localnet]` section. Update the program ID within this section:

    ```
    [programs.localnet]
    solana_custom = " 6xyMvrMKdjbXuTSjaDULm2vit8Uq6MjCgJj5N6zpJASY"
    ```

2. **lib.rs file of the Solana program**: Proceed to the `lib.rs` file of your Solana program, often located in `programs/solana-custom/src/lib.rs`. Update the program ID within this file as follows:

    ```
    use anchor_lang::prelude::*;
    use anchor_lang::solana_program::system_program;  declare_
    id!("6xyMvrMKdjbXuTSjaDULm2vit8Uq6MjCgJj5N6zpJASY");
    ```

3. After updating the program ID in these locations, the next crucial step involves rebuilding and redeploying your program to ensure the correct identifier is incorporated into your Solana project:

    ```
    anchor build
    anchor deploy
    ```

Updating the program ID ensures consistency and accuracy within your Solana project, allowing for seamless functionality and maintenance across various configurations and files.

Utilizing Anchor scripts

Let us now explore the utility of Anchor scripts, enabling streamlined execution of essential operations within your Solana project.

If you delve into your `Anchor.toml` file, you'll discover a dedicated `scripts` section containing predefined scripts, facilitating various actions within your project. One such script, specifically designed for running tests, is readily configured:

```
test = "yarn ts-mocha -p ./tsconfig.json -t 1000000 tests/**/*.ts"
```

This script is optimized to execute all tests located within the `tests` folder using Mocha, ensuring a comprehensive test suite for your Solana program.

Executing scripts with Anchor

To execute this predefined test script, simply run the following command:

```
anchor run test
```

If your local ledger is active via `solana-test-validator` and your project has been appropriately built and deployed using `anchor build and anchor deploy`, the executed tests should pass seamlessly!

Customizing Anchor scripts

The flexibility of Anchor scripts allows for the incorporation of custom commands within the `Anchor.toml` configuration file. For instance, you can add a custom script like this:

```
test = "..."
my-custom-script = "echo 'Hello world!'"
```

Executing the custom script is as simple as invoking the `anchor run` command:

```
anchor run my-custom-script
# Output: Hello world!
```

By utilizing these scripts, developers unlock a realm of possibilities—automating testing suites, performing routine tasks, or implementing specific functionalities within the Solana project environment. Their versatility and adaptability offer a robust framework for enhancing project efficiency and maintainability.

Exploration of Anchor scripts underscores their significance in streamlining operations, automating tasks, and fortifying the reliability of Solana programs. Let us now move on and test our dApp.

Testing your dApp

Ensuring your dApp works seamlessly and reliably is crucial before deployment. Testing helps verify the functionality and robustness of your Solana-based dApp. Anchor provides a comprehensive testing suite to simplify this process, allowing you to assess your dApp's behavior in a controlled environment.

To understand how testing fits into the development cycle, let us recap the necessary steps:

1. **Starting the local ledger**: Begin by initializing a local Solana ledger to simulate a Solana cluster within your local environment:

   ```
   solana-test-validator
   ```

2. **Building, deploying, and testing manually**: For manual build, deployment, and testing, you can do the following:

   - Use `anchor build` to compile the program.

   - Employ `anchor deploy` to deploy the program onto the local ledger.

   - Execute `anchor run test` to run tests specified in your `Anchor.toml` file.

   - However, Anchor offers a streamlined way to automate this cycle, making testing more efficient. Anchor introduces a dedicated command, `anchor test`, designed to handle the entire development cycle automatically. Let us break down how `anchor test` works.

3. **Initiating the local ledger**: The command starts a local ledger environment that is automatically terminated at the end of the testing process. It's crucial to note that if you have an active local ledger running, `anchor test` won't execute, so ensure to terminate any existing sessions beforehand:

   ```
   solana-test-validator -reset
   ```

4. **Building, deploying, and testing**: Once the local ledger is set up, Anchor proceeds with building the program, deploying it onto the local ledger, and executing the tests:

   ```
   anchor build
   anchor deploy
   anchor run test
   ```

Let's see what these commands achieve.

The `anchor test` command significantly streamlines the testing process. It abstracts away complexities, enabling you to focus on the actual functionality of your Solana program. However, there is a caveat when using `anchor test` after initializing a new project via `anchor init`. Running `anchor test` immediately after generating a new project via `anchor init` will not work until you update your program ID following the first deployment. Therefore, it is advisable to perform the initial build and deployment manually. After updating your program ID, you can start utilizing the efficiency of `anchor test`.

Testing your dApp using Anchor ensures a reliable and efficient development process, verifying the functionality and performance of your Solana-based applications. By leveraging `anchor test`, developers can automate the testing cycle, fostering a more focused and streamlined development journey.

This streamlined testing process significantly enhances the developer experience within the Solana ecosystem, ensuring a more reliable and robust final product.

Exploring localnet with Anchor

The `anchor localnet` command serves as a versatile tool within the Anchor framework, resembling `anchor test` in certain aspects. Unlike `anchor test`, this command does not execute any tests and doesn't terminate the local ledger after completion.

Essentially, running `anchor localnet` initiates a sequence of actions equivalent to the following:

```
solana-test-validator --reset
anchor build
anchor deploy
```

Once these steps are executed, the local ledger remains active, ensuring that your program persists for further development and interaction with your frontend client.

The key distinction between the two lies in the behavior after deployment—unlike `anchor test`, `anchor localnet` keeps the local ledger active. This retention of the ledger environment is valuable when you require continuous access to the deployed program for frontend development or other interactions.

Typically, developers utilize `anchor localnet` in scenarios where they predominantly focus on frontend client development, such as creating user interfaces or integrating the Solana-based functionality into web applications. This command streamlines the process of deploying the Solana program and maintains its availability for seamless integration with frontend components.

By retaining the active local ledger, `anchor localnet` offers an efficient workflow, enabling developers to swiftly iterate on their frontend interfaces while ensuring uninterrupted access to the deployed Solana program.

Moving on to the next section, let us see how to create accounts for our custom dApp.

Creating accounts for our custom dApp

In the Solana ecosystem, the principle that sets it apart lies in its account-based architecture. Unlike other blockchain platforms where smart contracts serve as data containers, Solana takes a unique stance—each piece of data necessitates its own dedicated account for storage. Accounts in Solana act as fundamental units for storing information, offering the flexibility for Solana programs to manage, manipulate, and interact with these accounts as required.

One of the most captivating aspects of Solana's architecture is the notion that programs themselves function as accounts. These program accounts possess distinctive characteristics—they house their code, remain immutable (read-only), and are designated as *executable*. A pivotal distinction within Solana's account structure is the inclusion of a Boolean value that identifies whether an account serves as a program account or a standard account intended for data storage. This foundational approach extends across diverse entities within Solana's ecosystem, encompassing wallets, **non-fungible tokens** (**NFTs**), tweets, and more—each represented and managed as an individual account. Consider this representation in *Figure 7.4*:

Account   Account   Executable Account
MESSAGE   USER   SOLANA
                                  CUSTOM PROGRAM

Figure 7.4 – Interconnectedness of accounts

This diagram portrays the interconnectedness of different accounts within Solana's framework. At the base level, a message is stored in an account, which is further linked to a user account. The user account, in turn, connects to an executable account identified as Solana Custom Program. This hierarchical visualization underscores the account-centric structure prevalent in Solana, where various entities and their interactions are encapsulated within distinct accounts. Let us now define accounts for our own dApp.

Defining accounts for our custom dApp

Before delving into the technicalities of account structuring, let's first understand the core functionality of our dApp. Our dApp enables users to send and receive messages in a decentralized manner, leveraging the power of blockchain technology for secure and verifiable communications. When creating accounts in Solana for a dApp, considerations around the structuring of accounts are crucial. Let us explore how accounts are typically defined for storing message-related information in a Solana program.

For our message-related functionality, structuring accounts appropriately is paramount. We will examine two possible solutions for storing messages:

- **Solution A (not scalable)**: This approach (*Figure 7.5*) involves consolidating all message-related data within a single account. While seemingly straightforward, this method imposes a critical limitation: the necessity to predetermine and allocate a fixed storage size capable of accommodating all potential messages. The fixed size allocation presents a scalability hurdle, especially in environments where the volume of messages fluctuates or expands unpredictably. The rigidity of this structure might lead to inefficiencies and constraints in managing a burgeoning number of messages. Consequently, the approach might become impractical as the application grows, potentially impeding scalability and flexibility:

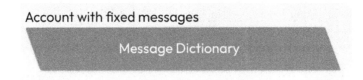

Figure 7.5 – Illustration of solution A

- **Solution B (scalable)**: In contrast, solution B (*Figure 7.6*) offers a highly efficient and scalable alternative by assigning individual accounts to each message. This architecture ensures that storage resources are generated and funded by the author of each message. By allocating dedicated accounts for every message, this solution optimizes storage utilization, allowing for a scalable and adaptable framework. Unlike solution A's fixed-size allocation, solution B's individual account model accommodates an unrestricted number of messages and users. Each message account operates independently, promoting efficient scalability and resource utilization, aligning well with the dynamic nature of dApps.

Figure 7.6 – Illustration of solution B

Now, let's see how to implement the message account structure.

Implementation of message account structure

In our Solana program, defining the **structure of the message account** follows a straightforward process. Within the lib.rs file, we employ a Message struct, encompassing vital properties essential for the message's representation within the Solana blockchain. These attributes include the following:

- author: Signifies the public key of the message's creator or sender
- timestamp: Indicates the precise time the message was posted or sent
- topic: An optional field allowing for the classification or categorization of messages, promoting better organization
- content: Stores the core content or body of the message itself

The usage of the #[account] attribute significantly simplifies the task of defining accounts within Solana programs. This attribute, supported by the Anchor framework, handles crucial operations such as account parsing and conversion from arrays of bytes. This abstraction mitigates the complexity involved in the intricate process of **account definition**, allowing readers to focus on the essential aspects of their dApp.

The inclusion of the author's public key within the account data serves as a fundamental aspect enabling specific user-based actions, such as message updates or deletions. This design choice empowers the dApp to facilitate author-specific interactions, ensuring distinct and personalized engagements within the dApp's ecosystem.

Understanding account sizing and rent in Solana

Within the Solana ecosystem, the sizing of accounts assumes considerable importance, particularly in light of the concept of **rent**. Rent, in the context of Solana, pertains to the cost attributed to the storage provided by an account. This fee is periodically collected by the blockchain. An account's inability to cover this rent poses a risk of deletion, potentially leading to the loss of associated data.

The creation of **rent-exempt accounts** involves allocating sufficient funds to cover a period of 2 years' worth of rent. By doing so, accounts become exempt from rent collection, ensuring their sustained existence without the looming threat of deletion. While Solana presents tools to calculate the minimum rent-exempt amount based on account size, the Anchor framework simplifies this process, automating the necessary calculations seamlessly.

Sizing message accounts

While designing a dApp on the Solana blockchain, particularly one that involves handling messages, careful consideration must be given to how data is structured and stored within the blockchain's accounts. Determining the size of message accounts necessitates evaluating the storage requirements of essential properties such as the author's public key, timestamp, topic, and content, taking into account their respective data types:

- **Author**: As a `PubKey` type, it occupies 32 bytes of storage

- **Timestamp**: Represented by the `i64` type, it consumes 8 bytes

- **Topic**: Defined as a `String` type with a maximum limit of 50 characters, it potentially requires 200 bytes along with a 4-byte prefix that indicates its length

- **Content**: Similarly, being a `String` type constrained to 280 characters, it may utilize 1,124 bytes with a 4-byte length prefix

These considerations regarding the sizes of individual message account properties facilitate efficient allocation of storage space, optimizing for the required capacity while avoiding unnecessary storage overheads.

Each piece of data stored needs to be meticulously planned in terms of its size and role within the account to ensure efficient use of storage and seamless functionality of the dApp. This includes accounting for various elements, such as account discriminators, the sizing of the author's public keys, timestamp storage, and the allocation of space for content and topics. Here's a detailed breakdown of these considerations:

- **Account discriminator**: In Solana, when creating a new account, an 8-byte **discriminator** is added at the beginning of the data. This discriminator serves as a type identifier for the account, distinguishing different account types within the program. As we establish the size of our message account, we must consider this 8-byte discriminator, essential for recognizing the tweet account within the program's storage.

- **Sizing the author's public key**: The author's public key, represented by the PubKey data type, demands 32 bytes of storage. This key is vital as it links each message to its respective author, allowing for proper attribution and author-specific actions within the dApp.

- **Timestamp storage**: For the timestamp property, an i64 type that represents an integer of 64 bits or 8 bytes suffices. Storing the time of message publication in this format ensures accurate time tracking within the program.

- **Determining topic size**: The topic field, characterized as a String type, presents a nuanced challenge in sizing. By setting a maximum limit of 50 characters for a topic, we tentatively allocate 200 bytes, which is 4 bytes per character, for its storage. It's important to note that this allocation accounts for a maximum scenario based on the chosen character encoding.

- **Content storage considerations**: Similar to the topic, the content field, also a String type, is constrained to a maximum of 280 characters, corresponding to 1124 bytes of storage. This allocation, in combination with a 4-byte length prefix, ensures efficient handling of message content, accommodating variations in character lengths.

Now, we will implement the account in code.

Implementation in code

The lib.rs file encapsulates the structuring of the message account and defines essential constants that detail the byte sizes of various account properties:

```
#[account]
pub struct Message {
    pub author: Pubkey,
    pub timestamp: i64,
    pub topic: String,
    pub content: String,
}

// Constants defining sizes of account properties
const DISCRIMINATOR_LENGTH: usize = 8;
const PUBLIC_KEY_LENGTH: usize = 32;
```

```
const TIMESTAMP_LENGTH: usize = 8;
const STRING_LENGTH_PREFIX: usize = 4; // Stores the size of the
string.
const MAX_TOPIC_LENGTH: usize = 50 * 4; // 50 chars max.
const MAX_CONTENT_LENGTH: usize = 280 * 4; // 280 chars max.

impl Message {
    // Constant representing the total size of the Message account
    const LEN: usize = DISCRIMINATOR_LENGTH
        + PUBLIC_KEY_LENGTH // Author.
        + TIMESTAMP_LENGTH // Timestamp.
        + STRING_LENGTH_PREFIX + MAX_TOPIC_LENGTH // Topic.
        + STRING_LENGTH_PREFIX + MAX_CONTENT_LENGTH; // Content.
}
```

We will now add this code to the existing file. In this code segment, the LEN constant within the impl Message block aggregates the sizes of all properties, providing the total byte size for a single message account. This consolidated size offers a quick reference to developers and aids in the accurate allocation of account storage within the Solana blockchain.

Next, we will jump on to the next section and understand how to create our first instruction for our custom dApp.

Creating our first instruction

As we proceed into the next phase of our Solana-based dApp development, our journey so far has been a meticulous exploration of architecting message accounts. We've meticulously established account structures, gauged sizes, and delved into rent concepts, all to fortify the groundwork for our dApp's functionality. Now, the focus shifts toward the pivotal juncture of creating our initial instruction. At this juncture, our focus spans various critical facets: setting account boundaries, translating logic into code, fortifying against invalid inputs, and understanding the nuanced difference between instructions and transactions. Each step within this undertaking revolves around the core aspect of message accounts, all geared toward building a resilient, efficient, and robust dApp core within Solana's dynamic environment.

Let us start with an understanding of how we will create instructions.

Introduction to instruction creation

In Solana's decentralized ecosystem, creating instructions is key to triggering actions on the network, a step we'll explore as we navigate its account-centric architecture. Solana's stateless programs necessitate a comprehensive inclusion of all requisite contexts for these instructions to execute seamlessly.

In our pursuit to enable essential functionalities within our dApp, the structuring of contexts is paramount. Much like the structural definition of our `Message` account, contexts are crafted through the implementation of a Rust struct. This struct encapsulates essential accounts vital for an instruction to execute effectively within our Solana program.

Upon revisiting the `lib.rs` file, above the previously defined `Message` struct, you'll likely encounter an empty `Initialize` context, resembling the following:

```
#[derive(Accounts)]
pub struct Initialize {}
```

In this instance, the `Initialize` context serves as a mere placeholder, devoid of the accounts necessary for an instruction's execution. Replacing this placeholder context with a more purposeful and comprehensive structure named `SendMessage` is essential. The `SendMessage` context becomes the backbone for our instruction, defining crucial accounts imperative for executing this specific instruction.

To achieve this, simply remove the existing `Initialize` context and supplant it with the following code block:

```
#[derive(Accounts)]
pub struct SendMessage<'info> {
    pub message: Account<'info, Message>,
    pub author: Signer<'info>,
    pub system_program: AccountInfo<'info>,
}
```

Breaking down the `SendMessage` context, we find three fundamental accounts explicitly enlisted to ensure the seamless execution of our `SendMessage` instruction:

- `message`: Represents the account responsible for storing the message, necessitating an association with the instruction

- `author`: Denotes the signer required to validate the authority and authenticity of the message sender

- `system_program`: Refers to the official System Program provided by Solana, critical for initializing the `Message` account and assessing the requisite rent-exempt status

Upon introducing the `SendMessage` context comprising specific accounts required for executing our Solana instruction, a few novel concepts and Rust features come into play. Firstly, the `SendMessage` context encapsulates key accounts crucial for the successful execution of the `SendMessage` instruction within our Solana program.

Let's delve into the distinctive accounts specified within the `SendMessage` context and their significance:

- `message`: This account embodies the foundational element responsible for storing the message. Despite this instruction being the initiator of the account creation, providing the `message` account here entails passing the public key for the account's creation. The provision of this key facilitates the instruction in the creation of the `Message` account. Additionally, signing the instruction using its private key verifies ownership, essentially instructing, *"Here's my public key; kindly create a Message account for me."*

- `author`: Emphasizing the essence of authentication, this account ensures the identification of the sender initiating the message. Requiring the author's signature assures the authenticity of the sender's identity, preventing any attempt at transmitting a message on behalf of someone else.

- `system_program`: Central to the functionality due to Solana's stateless nature, this account represents Solana's official System Program. Despite programs being stateless, `system_program` becomes integral for initializing the `Message` account and determining the necessary rent-exempt status.

Now, let's decipher some Rust-specific intricacies evident in the aforementioned `SendMessage` code snippet:

- `#[derive(Accounts)]`: This attribute, a product of Anchor, simplifies the generation of code and macros for the `SendMessage` context struct. Its presence significantly streamlines otherwise complex lines of code, ensuring a more comprehensible struct definition.

- `<'info>`: This Rust lifetime, though defined as a generic type, primarily informs the Rust compiler about the duration of variable lifetimes. It aids in clarifying how long specific variables remain active within the program, ensuring proper memory management.

In terms of types represented within the context, we have the following:

- `AccountInfo`: Acting as a fundamental Solana structure, `AccountInfo` serves as a low-level representation capable of accommodating any account. When utilizing `AccountInfo`, the account's data manifests as an unparsed array of bytes.

- `Account`: An Anchor-provided account type that encapsulates `AccountInfo` within another struct, enabling data parsing as per the specified account struct (in this case, `Message`). The `Account<'info, Message>` designation explicitly denotes an account of type `Message`, parsing data accordingly.

- `Signer`: Similar to `AccountInfo`, `Signer` represents an account type but with the additional requirement of signing the instruction, ensuring validation through the provided signature.

These Rust features, while seemingly intricate, play pivotal roles in ensuring efficient structuring and execution of instructions within the Solana ecosystem. Additionally, employing account constraints enables specific assertions, ensuring the desired properties of various account types. Let us look at these account constraints next.

Establishing account constraints

Establishing **account constraints** within a Solana program is akin to setting rules of engagement for different account types. In this phase of our dApp development, we delve into defining specific guidelines and constraints for accounts, enabling a meticulous control mechanism for how these accounts interact within the program. Account constraints are crucial for security in Solana's dApps. They make sure each account is used correctly, whether it's signing transactions, managing data, or using features. By setting these rules, developers create a secure and reliable environment for their dApps to operate on the Solana blockchain This section will intricately explore the nuances of these constraints, shedding light on their significance, implementation, and impact on the overall functionality of our dApp.

When defining a new account within our instruction—specifically for our message—implementing constraints ensures a streamlined process for its initiation. To integrate these constraints effectively, incorporate the following line atop the `message` property:

```
pub struct SendMessage<'info> {
    #[account(init, payer = author, space = 500, seeds =
[&[b"message"[..]], &[author.key.as_ref()]], bump = 9)]
    pub message: Account<'info, Message>,
    pub author: Signer<'info>,
    pub system_program: AccountInfo<'info>,
}
```

However, a straightforward addition such as this might trigger an error within the code structure. This error stems from the absence of crucial information—specifically regarding the storage space required for our message account and determining the entity responsible for funding the rent-exempt allocation. Leveraging the `payer` and `space` arguments significantly streamlines our code for handling message accounts within the instruction:

```
#[derive(Accounts)]
pub struct SendMessage<'info> {
    #[account(init, payer = author, space = Message::LEN)]
    pub message: Account<'info, Message>,
    pub author: Signer<'info>,
    pub system_program: AccountInfo<'info>,
}
```

These arguments wield immense power within a single line of code, referencing vital elements crucial for the successful execution of our Solana-based dApp functionalities. The `payer` argument seamlessly ties to the author's account within the same context. Simultaneously, the `space` argument efficiently allocates storage by utilizing the previously defined `Message::LEN` constant, simplifying the account initialization process remarkably.

With the declaration that the author bears the cost for the rent-exempt allocation of the message account, it becomes essential to mark the author's property as mutable due to the need for potential alterations in their account balance. Anchor simplifies this process using the mut account constraint, facilitating the necessary account mutation effortlessly:

```
#[derive(Accounts)]
pub struct SendMessage<'info> {
    #[account(init, payer = author, space = Message::LEN)]
    pub message: Account<'info, Message>,
    #[account(mut)]
    pub author: Signer<'info>,
    pub system_program: AccountInfo<'info>,
}
```

The mut constraint signifies the mutable nature of the author's account, allowing adjustments in its state during the process. Additionally, though a signer account constraint could validate the author's signature, it's redundant in this context as the Signer account type inherently ensures the signing attribute.

It's crucial to constrain the system_program account to guarantee its legitimacy as the official System Program from Solana. Without this constraint, there's a risk of potential malicious actors supplying unauthorized system programs, compromising the integrity of the transaction:

```
#[derive(Accounts)]
pub struct SendMessage<'info> {
    #[account(init, payer = author, space = Message::LEN)]
    pub message: Account<'info, Message>,
    #[account(mut)]
    pub author: Signer<'info>,
    #[account(address = system_program::ID)]
    pub system_program: AccountInfo<'info>,
}
```

Utilizing the address account constraint with system_program::ID ensures an exact match between the provided public key and the defined constant in Solana's code base. However, note that system_program::ID isn't included in Anchor's prelude::*, necessitating an additional line at the top of the lib.rs file to import this constant:

```
use anchor_lang::prelude::*;
use anchor_lang::solana_program::system_program;
```

In the latest iterations of Anchor, there's an alternative method to validate the System Program's authenticity using a distinct account type named Program. By employing this account type and providing it with the System type, we ascertain that it specifically refers to the official System Program:

```
#[derive(Accounts)]
pub struct SendMessage<'info> {
```

```
        // … other account declarations
    }
```

This streamlined approach offers a concise way to establish the context for our `SendMessage` instruction, simplifying the process of verifying the System Program's validity.

Let us now understand in the next section how we implement logic.

Implementing logic

As we step into the heart of our Solana-based dApp development, the implementation of logic within our instruction takes center stage. This pivotal phase involves translating the conceptual architecture we've meticulously constructed into tangible code. Our focus now shifts from defining structures and constraints to breathing life into our application's functionality. With a keen eye on the intricacies of Solana's programming paradigm, we embark on the journey of encoding core operations for sending a message within our dApp. Each line of code etches a pathway toward robust functionality, laying the groundwork for a resilient and efficient dApp within the dynamic Solana ecosystem.

The `solana_custom` module in the following code snippet is utilized to encapsulate our message-sending functionality. It's important to note that this module is defined within the `lib.rs` file of our project, which serves as the entry point for the Rust library:

```
#[program]
pub mod solana_custom {
    use super::*;
        // CODE
    }
}
```

In our `solana_custom` module, let's transition from the initialization phase to the implementation of our `SendMessage` instruction by revising the initial function. Replace the existing `Initialize` function with the following code:

```
pub fn send_message(ctx: Context<SendMessage>, topic: String, content:
String) -> ProgramResult {
    Ok(())
}
```

This code represents a pivotal shift in our application's flow. Several notable alterations are introduced:

- **Renaming**: The `initialize` function has been rebranded to `send_message()`. Rust follows the snake case for function names.
- **Context association**: We've specified the generic type inside `Context` as `SendMessage`, establishing a direct link between the instruction and the predefined context.

- **Additional arguments**: We introduced two new arguments, `topic` and `content`, essential components encapsulating the essence of the message. These arguments, unlike accounts, are parameters passed directly to the function.

The function returns a `ProgramResult` instance, a Rust enum signaling either `Ok` (successful execution) or `ProgramError` (execution failure). Rust foregoes exceptions, relying on this special enum to communicate execution status. For now, the function returns `Ok(())` —an `Ok` type with an empty value inside `()` —signifying an uneventful yet successful execution.

To proceed, let's access the message account from the context. Thanks to Anchor's `init` account constraint acting as middleware, the necessary accounts are prepared before the function execution. Let's walk through how we'd approach this process:

```
pub fn send_message(ctx: Context<SendMessage>, topic: String, content:
String) -> ProgramResult {
    // Accessing the message account from the context
    let message_account: &mut Account<Message> = &mut ctx.accounts.
message;

    // Additional logic can be implemented here
    // ...

    Ok(())
}
```

By referencing `ctx.accounts.message`, annotated with &mut to signify mutable borrowing, we obtain a mutable reference to the `message` account. This reference, stored in the `message_account` variable, enables us to modify or interact with the account's data within the function.

Similarly, when retrieving the author information to link it with the message account, we access the `author` account through the context. Here's how we'd incorporate this:

```
pub fn send_message(ctx: Context<SendMessage>, topic: String,
content: String) -> ProgramResult {
    let message_account: &mut Account<Message> = &mut ctx.
accounts.message;
    let author: &Signer = &ctx.accounts.author;

    // Further operations involving author or message_account
    // ...

    Ok(())
}
```

The `author` account is accessed via `ctx.accounts.author`, annotated simply with & as it's a reference and doesn't require mutability since the rent-exempt payment for this account is managed by Anchor. This provides us with the necessary data to establish associations or perform specific actions linked to the author within the context of the `send_message()` function.

Fetching the system's current timestamp from Solana's `Clock` system variable is crucial for our message creation process. By incorporating the `Clock` system variable and ensuring the presence of the System Program account, we can successfully gather the necessary timestamp for the message. Let's go over the code snippet that accomplishes this task:

```
pub fn send_message(ctx: Context<SendMessage>, topic: String,
content: String) -> ProgramResult {
    let message_account: &mut Account<Message> = &mut ctx.
accounts.message;
    let author: &Signer = &ctx.accounts.author;
    let clock: Clock = Clock::get().unwrap();

    // Additional logic involving clock timestamp assignment to the
message_account
    // ...

    Ok(())
}
```

Here, the `Clock::get()` function retrieves the current system time, and by utilizing `unwrap()`, we handle the `Result` type that `Clock::get()` returns. Unwrapping the result enables us to access the timestamp data provided by `Clock::get()` and proceed with our message creation process.

By combining the gathered timestamp with the provided topic and content, we've gathered all the essential data required to populate our new message account effectively. This data aggregation sets the stage for the subsequent steps involved in creating a robust and complete message within our Solana program.

To craft a comprehensive message creation process, we need to meticulously fill in the essential details. Starting with the author's public key, we access it via `author.key`. However, since it's a reference, we dereference it using * to access the actual public key data:

```
pub fn send_message(ctx: Context<SendMessage>, topic: String, content:
String) -> ProgramResult {
    let message_account: &mut Account<Message> = &mut ctx.accounts.
message;
    let author: &Signer = &ctx.accounts.author;
    let clock: Clock = Clock::get().unwrap();

    message_account.author = *author.key;
    // Additional logic involving clock timestamp assignment to the
```

```
message_account
    // ...

    Ok(())
}
```

Next, retrieving the Unix timestamp from the system clock is done using `clock.unix_timestamp`:

```
pub fn send_message(ctx: Context<SendMessage>, topic: String, content:
String) -> ProgramResult {
    let message_account: &mut Account<Message> = &mut ctx.accounts.
message;
    let author: &Signer = &ctx.accounts.author;
    let clock: Clock = Clock::get().unwrap();

    message_account.author = *author.key;
    message_account.timestamp = clock.unix_timestamp;
    // Additional logic involving topic and content assignment to the
message_account
    // ...

    Ok(())
}
```

Lastly, storing the provided `topic` and `content` parameters in their respective properties within the `message_account` variable completes the process:

```
pub fn send_message(ctx: Context<SendMessage>, topic: String, content:
String) -> ProgramResult {
    let message_account: &mut Account<Message> = &mut ctx.accounts.
message;
    let author: &Signer = &ctx.accounts.author;
    let clock: Clock = Clock::get().unwrap();

    message_account.author = *author.key;
    message_account.timestamp = clock.unix_timestamp;
    message_account.topic = topic;
    message_account.content = content;

    Ok(())
}
```

This comprehensive process ensures that a message account is meticulously created and equipped with the correct information, setting the foundation for a robust message creation mechanism within our Solana program.

Safeguarding against invalid data

Safeguarding against invalid data is a critical aspect of any dApp built on Solana. In the realm of blockchain, ensuring the integrity and validity of data is paramount, given its immutable nature once recorded. In Solana's decentralized ecosystem, where every piece of information resides within an account, maintaining the accuracy and authenticity of data becomes even more crucial. Invalid or corrupted data can disrupt the entire system, affecting functionalities, compromising security, and eroding user trust. Hence, implementing robust measures to safeguard against invalid data is a fundamental pillar in developing reliable, secure, and resilient dApps on the Solana blockchain. This entails employing stringent validation mechanisms, ensuring data integrity during processing and storage, and integrating fail-safes to rectify or prevent erroneous data from impacting the system's performance and overall functionality.

Ensuring the validity and integrity of data within our Solana program is crucial for maintaining system reliability. While Anchor's account constraints provide a level of protection against certain invalid scenarios, we need to implement additional checks to enforce our specific data requirements.

In our program, we've designated the `topic` and `content` properties as `String` types, allocating 50 characters for the former and 280 characters for the latter. However, as the `String` type doesn't inherently limit the number of characters, there's a possibility for users to exceed these limits, potentially causing issues. For instance, a user could input a topic of 280 characters and content of 50 characters, or even content that's (280 + 50) * 4 = 1,320 characters long, utilizing the `String` type's flexible nature.

To prevent such occurrences and ensure compliance with our predefined constraints, we're implementing safeguards within our code. By introducing conditional checks using `if` statements before hydrating our message account, we'll verify that both the `topic` and `content` parameters adhere to our character limits. Utilizing the `chars().count()` method enables us to accurately assess the number of characters within a `String` type, as opposed to the byte count provided by the `len()` method associated with vectors.

Here's a snippet of code showcasing these validations:

```
pub fn send_message(ctx: Context<SendMessage>, topic: String, content:
String) -> ProgramResult {
    let message: &mut Account<Message> = &mut ctx.accounts.message;
    let sender: &Signer = &ctx.accounts.sender;
    let clock: Clock = Clock::get().unwrap();

    if topic.chars().count() > 50 {
        // Return an error to handle the exceeding topic character
    limit...
    }

    if content.chars().count() > 280 {
        // Return an error to handle the exceeding content character
```

```
limit...
    }

    message.sender = *sender.key;
    message.timestamp = clock.unix_timestamp;
    message.topic = topic;
    message.content = content;
    Ok(())
}
```

These conditional statements, when triggered, will halt the execution of the instruction and return an error, effectively preventing the acceptance of invalid data that doesn't align with our specified constraints. This meticulous validation process ensures the accuracy and compliance of incoming data, fortifying the robustness of our Solana dApp against potential data integrity issues.

Anchor simplifies error handling by enabling the definition of an `ErrorCode` enum using the `#[error_code]` Rust attribute. This enum allows us to categorize various errors within our Solana program and associate each error type with a descriptive message using the `#[msg("...")]` attribute.

Let's integrate our custom `ErrorCode` enum into the existing code base. By defining two specific errors—one for cases when the topic exceeds the character limit and another for exceeding the content's character limit—we establish a clear framework for handling these scenarios.

Here's the `ErrorCode` enum, along with its designated errors placed at the end of the `lib.rs` file:

```
#[error_code]
pub enum ErrorCode {
    #[msg("The provided topic should be 50 characters long maximum.")]
    TopicTooLong,
    #[msg("The provided content should be 280 characters long
maximum.")]
    ContentTooLong,
}
```

Implementing these errors into our message-sending logic involves utilizing conditional checks to verify the length of the `topic` and `content` parameters. If either parameter exceeds its predefined character limit, the respective error from our `ErrorCode` enum will be returned using the `Err` variant, effectively halting the instruction and indicating the specific error:

```
anchor_lang::solana_program::entrypoint::ProgramResult;
pub fn send_message(ctx: Context<SendMessage>, topic: String, content:
String) -> ProgramResult {
    let message: &mut Account<Message> = &mut ctx.accounts.message;
    let sender: &Signer = &ctx.accounts.author;
    let clock: Clock = Clock::get().unwrap();
    if topic.chars().count() > 50 {
```

```
            return Err(error!(ErrorCode::TopicTooLong).into());
        }
        if content.chars().count() > 280 {
            return Err(error!(ErrorCode::ContentTooLong).into()
        }
        Message.author= *author.key;
        message.timestamp = clock.unix_timestamp;
        message.topic = topic;
        message.content = content;
        Ok(())
    }
```

By incorporating these error-handling mechanisms, we fortify our Solana dApp against potential issues stemming from invalid data input, ensuring stringent adherence to our predefined constraints. Furthermore, this setup allows easy scalability—additional error types and guards can be seamlessly added and expanded upon as our program evolves.

Instruction versus transaction

Understanding nuances between **instructions** and **transactions** within the Solana blockchain ecosystem is fundamental to developing efficient and effective dApps. In Solana, these terms represent distinct components that contribute to the execution and management of operations on the blockchain. Instructions and transactions serve unique purposes and play vital roles in orchestrating decentralized processes while adhering to Solana's high-performance and scalable architecture.

Instructions serve as fundamental units of operations within Solana's smart contracts or programs. These instructions encapsulate specific actions or tasks that are executed by Solana's validators across the network. They embody granular commands such as transferring tokens, updating data, or triggering specific functions within smart contracts. Each instruction carries out a single operation and contains information and parameters necessary to perform the intended action. Consequently, a single transaction in Solana can encompass multiple instructions, allowing for efficient batch processing and the execution of various operations within a single atomic transaction.

On the other hand, transactions in Solana act as containers or bundles that group multiple instructions together, forming a cohesive set of actions to be executed on the blockchain. These transactions serve as the means to atomically execute a sequence of instructions, ensuring that either all operations within the transaction succeed or none of them do. Transactions also include essential metadata such as signatures, timestamps, and fee information necessary for validation and processing by Solana's network.

Distinguishing between instructions and transactions is crucial for dApp developers to optimize the execution of tasks on the Solana blockchain. Understanding how instructions encapsulate specific actions and how transactions bundle these instructions for atomic execution allows for the creation of sophisticated and efficient dApps. By leveraging the flexibility and scalability inherent in Solana's architecture, developers can design dApps that achieve precise and complex functionalities while maintaining the integrity and security of operations within the blockchain network.

Moving on, we will now understand how to create tests for our instructions.

Creating tests for our instructions

Testing the functionality and integrity of instructions within a Solana-based dApp is a critical step in ensuring the reliability and robustness of smart contracts or programs. As developers craft intricate instructions to execute specific tasks on the blockchain, validating these instructions through comprehensive testing becomes indispensable. The testing process involves creating scenarios that mimic real-world interactions with the dApp, verifying that instructions execute as intended and handle various edge cases or unexpected behaviors gracefully. Through systematic and thorough testing, developers can identify and rectify potential bugs, vulnerabilities, or inefficiencies in their smart contracts, ensuring the stability and security of the dApp once deployed onto the Solana blockchain.

Testing instructions involves simulating diverse scenarios, such as valid and invalid inputs, exceptional conditions, and edge cases, to validate the correctness and resilience of instructions in varying circumstances. By designing test cases that encompass a wide range of potential interactions, developers can ascertain that instructions perform accurately, adhere to predefined logic, and respond appropriately to unexpected situations. Additionally, comprehensive testing aids in benchmarking the performance of instructions, ensuring they meet expected throughput and latency requirements within the Solana network.

Delving into the world of testing might seem less glamorous, but it's a gateway to understanding our program's interaction mechanisms and user experiences. While our focus so far has been on developing a blockchain-based program, it's time to venture to the other side—the client interface.

This transition introduces us to the realm of client-side operations, akin to traditional web server-client interactions. To test our Solana program, we'll utilize a JavaScript client temporarily. This serves as an invaluable bridge between our blockchain program and the forthcoming Vue.js based JavaScript client, allowing us to perfect our interactions with the program before deploying the frontend.

Interacting with the Solana blockchain involves leveraging its JSON RPC API. Despite the technical nomenclature, it essentially functions as an API enabling communication. Simplifying this process, Solana offers the `@solana/web3.js` JavaScript library, which streamlines interaction with the RPC API. This library furnishes an array of asynchronous methods nested within a `Connection` object. This object operates within a cluster, defining the destination for requests—be it `localhost`, `devnet`, or any other designated cluster. To launch our dApp to blockchain we need to first test it on localhost. For the local testing purposes, we'll employ the `localhost` cluster, initiating our program to blockchain.

Let's talk briefly talk about the clusters with evolving visual representation. The initial diagram (*Figure 7.7*) illustrates two key nodes—an overarching **Cluster node** connected to a nested **Connection node**.

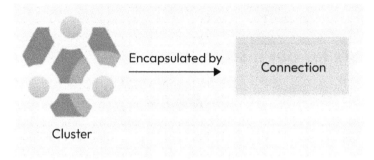

Figure 7.7 – Illustration of a cluster

Understanding how to communicate with the Solana blockchain is one facet, but authenticating transactions requires unique identity verification. This validation is achieved through the utilization of a `Wallet` object, essential for accessing the user's key pair—a fundamental element in transaction signing.

Thankfully, the convenience doesn't end there. Anchor extends its support to the client side with its `@project-serum/anchor` JavaScript library, simplifying this process further. This library introduces a `Wallet` object that streamlines transaction signing by incorporating the necessary key pair. Beyond that, it provides a `Provider` object, amalgamating both the connection and wallet functionalities. Seamlessly integrated, this `Provider` object ensures that the wallet's signature is automatically included in outbound transactions, making interactions with the Solana blockchain on behalf of a user's wallet smooth and effortless.

Figure 7.8 visually reinforces this interconnection: the **Cluster** node linked to the **Connection** node signifies the foundational communication structure, while the subsequent link from the **Wallet** node to the **Provider** node indicates the unified functionality encapsulated within the `Provider` object:

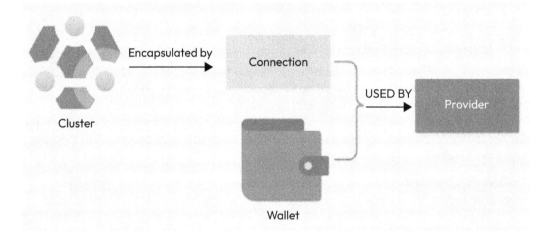

Figure 7.8 – Illustration of the usage of wallet and connection

Remember that handy IDL file named `idl/` generated by Anchor whenever we run `anchor build`? That file is a goldmine of structured program details—it encapsulates our program's public key, instructions, and accounts in a neatly organized format. Now, consider merging this IDL file, packed with comprehensive program knowledge, with the `Provider` object we discussed earlier—the one that effortlessly communicates with the Solana blockchain on behalf of a wallet. Imagine the synergy!

Here's where Anchor delivers yet another game-changer: the `Program` object. This nifty addition utilizes both the IDL file and the `Provider` object. It crafts a tailored JavaScript API, mirroring every aspect of our Solana program. With this `Program` object, interacting with our Solana program becomes a breeze, even on behalf of a wallet. No need to delve into the underlying API intricacies—it's all streamlined (*Figure 7.9*):

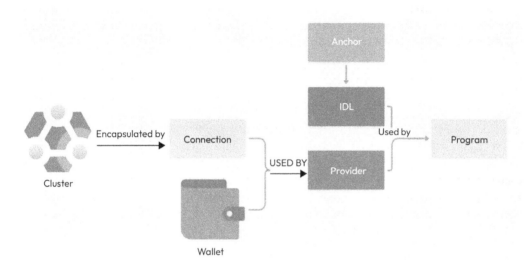

Figure 7.9 – Developer-centric experiences in Solana development

This amalgamation of the IDL file's insights and the `Provider` object's functionality empowers developers to effortlessly engage with their Solana programs, all thanks to Anchor's innovative approach. It's the final piece completing the puzzle of developer-centric experiences in Solana development.

Creating a client for tests

Let's dive into practical implementation by setting up a `Program` object for use in our tests. No need to import additional JavaScript libraries—both necessary libraries, `@solana/web3.js` and `@project-serum/anchor`, come pre-packaged in every Anchor project. Referring to the earlier diagram, two critical questions stand: which cluster and which wallet to utilize? Anchor simplifies this process by generating a `Provider` object, leveraging the configurations specified within our `Anchor.toml` file.

Specifically, it looks for your provider configurations, which typically resemble the following:

```toml
[provider]
cluster = "localnet"
wallet = "/Users/{your username}/.config/solana/id.json"
```

These configurations serve as directives for the `Provider` object. For instance, it automatically selects the `localhost` cluster, making use of a local ledger. Additionally, it knows precisely where to locate your key pair on your local machine.

This seamless configuration handling by Anchor ensures that our `Provider` object is equipped with the necessary settings, paving the way for effortless interaction with our Solana program. Navigate to the test file, typically located at `tests/solana-custom.ts`. Inside the `describe()` method, you'll encounter the following initial lines of code:

```
// Configure the client to use the local cluster.
anchor.setProvider(anchor.Provider.env());
const program = anchor.workspace.SolanaMessage as
Program<SolanaMessage>;
```

The first line is a call to `anchor.Provider.env()`, a method that generates a new `Provider` object using the configurations specified in our `Anchor.toml` file. This process follows this simple formula: `Cluster + Wallet = Provider`. Subsequently, this newly created provider is registered using the `anchor.setProvider` method, ensuring that our tests will interact with the specified Solana network environment.

Moving to the second line, it employs the registered provider to create a new `Program` object. In TypeScript, this object is tied to the specific `SolanaCustom` program generated by Anchor during the anchor build process. This linkage to the `SolanaCustom` program type enhances the development experience by enabling useful autocompletion features within code editors. And just like that, our test client is all set up and ready to be used!

Sending a message

Start by removing the default test template in the `tests/solana-custom.ts` file. Replace it with the following code snippet:

```
it('can send a new message', async () => {
    await program.rpc.sendMessage('TOPIC HERE', 'CONTENT HERE', {
        accounts: {
            // Accounts here...
        },
        signers: [
            // Key pairs of signers here...
```

```
    ],
  });
});
```

Now, let's break down this test scenario:

1. **Test setup**: We've initiated a new test named `can send a new message` using Mocha's `it` method, which is ideal for creating individual test cases.

2. **Asynchronous function**: This test function is marked as `async` because it involves asynchronous operations. Specifically, we'll await the completion of the transaction before proceeding to verify the outcome.

3. **Interacting with the program**: Using the `Program` object, which encapsulates our Solana program, we access the `rpc` object. This `rpc` object mirrors our program's instructions. In this case, to trigger the `SendMessage` instruction, we call `program.rpc.sendMessage`.

4. **Providing instruction arguments**: The `sendMessage` method requires two arguments: the topic and content of the message. Replace `'TOPIC HERE'` and `'CONTENT HERE'` with your desired topic and content strings.

5. **Context and signers**: The last argument of the `program.rpc` method is the context. This object contains all necessary accounts for the instruction to execute successfully. Ensure to provide the required accounts and their corresponding key pairs in the `accounts` and `signers` fields, respectively. Note that the wallet's key pair isn't explicitly provided, as Anchor handles this automatically.

You might notice TypeScript flagging errors due to incomplete data. We'll gradually add the required data to address these issues and progress with the test implementation. Let us fill in the `sendMessage` method:

```
it('can send a new message', async () => {
    // Generate a new key pair for the message account (message).
    const message = anchor.web3.Keypair.generate();

    // Send a message with a space theme (topic: space exploration,
content: Discovering new worlds!)
    await program.rpc.sendMessage('space exploration', 'Discovering
new worlds!', {
        accounts: {
            message: message.publicKey, // Define the message account.
        },
        signers: [message], // Add the message key pair as a signer.
    });
```

This test scenario simulates the sending of a new message within the Solana program, specifically themed around space exploration with the `Discovering new worlds!` content:

```
it('can send a new message', async () => {
    // Test code here...
});
```

This test function, written using the `it` function from the Mocha test framework, checks the ability of the program to send a new message:

```
const message = anchor.web3.Keypair.generate();
```

The `anchor.web3.Keypair.generate()` method creates a new key pair (`message`), which consists of a public key and a corresponding private key. This key pair will be associated with the new message account:

```
await program.rpc.sendMessage('space exploration', 'Discovering new
worlds!', {
    accounts: {
        message: message.publicKey, // Define the message account.
    },
    signers: [message], // Add the message key pair as a signer.
});
```

Let's discuss this method in detail:

- The `program.rpc.sendMessage()` method invokes the `sendMessage` instruction in the Solana program. This instruction is designed to send a new message.

- The method takes three parameters:

  - `space exploration` represents the topic of the message, indicating a theme related to space exploration.

  - `Discovering new worlds!` is the content of the message.

  - The third parameter is an object containing the following:

    - `accounts` field: Defines the accounts involved in the instruction. Here, it specifies the message account created earlier.

    - `signers` field: Provides the necessary signers for executing this instruction. In this case, `message` (the key pair) is added as a signer.

As Anchor automatically includes the wallet as a signer for every transaction, there's no need for us to modify the `signers` array. Lastly, we must supply the `system_program` account. In JavaScript, Anchor automatically converts snake-case variables to camel-case within our context. This means we need to reference the System Program using `systemProgram` instead of `system_program`.

Accessing Solana's official System Program's public key in JavaScript is straightforward. We utilize `anchor.web3.SystemProgram` to access the System Program and retrieve its public key through `anchor.web3.SystemProgram.programId`:

```
it('can send a new message', async () => {
    // Generate a new key pair for the message account.
    const message = anchor.web3.Keypair.generate();

    // Send a message about "space exploration" with content
"Discovering new worlds!".
    await program.rpc.sendMessage('space exploration', 'Discovering
new worlds!', {
        accounts: {
            message: message.publicKey, // Define the message account.
            author: program.provider.wallet.publicKey, // Identify the
author using the wallet's public key.
            systemProgram: anchor.web3.SystemProgram.programId, //
Provide the System Program.
        },
        signers: [message], // Add the message key pair as a signer.
    });
});
```

We're now set up to send messages to our Solana program. However, simply running this test doesn't validate much. We need to interact further by fetching the newly created account on the blockchain and verifying the data matches what we have sent.

To retrieve an account from the blockchain, we utilize additional APIs provided by the `Program` object. Using `program.account.message`, we gain access to specific methods designed to fetch message accounts from the blockchain. Keep in mind that these methods are available for each account defined within our Solana program. For instance, if we had a `UserProfile` account, we'd access it using the `program.account.userProfile` API.

Within these API methods, we employ `fetch` to precisely retrieve a single account by supplying its public key. Anchor comprehends the type of account we are fetching and automatically handles the data parsing for us.

Let's retrieve our recently created `Message` account. We'll utilize the public key associated with our message key pair to fetch `messageAccount`. Additionally, let's log the contents of that account to inspect its details:

```
it('can send a new message', async () => {
    const message = anchor.web3.Keypair.generate();
    await program.rpc.sendMessage('space exploration', 'Discovering
new worlds!', {
        ...
```

```
    });
    const messageAccount = await program.account. message.fetch(message.
    publicKey);
        console.log(messageAccount);
    });
```

Now, let's execute our tests by using the `anchor test` command. This command is responsible for building, deploying, and testing our program on its local ledger. For Apple M1 users, it's essential to run two commands separately, `solana-test-validator --no-bpf-jit --reset` and `anchor test --skip-local-validator`, ensuring smooth execution. Upon running the tests, you'll notice the expected behavior of the tests passing, which is normal since we haven't set up any specific assertions yet. However, along with the passing tests, you'll encounter an object logged in the console resembling the following structure:

```
    author: PublicKey {
      _bn: <BN: 6xyMvrMKdjbXuTSjaDULm2vit8Uq6MjCgJj5N6zpJASY6xyMvr>
    },
    timestamp: <BN: 1a2b3c4d>,
    topic: 'space exploration',
    content: 'Discovering new worlds!'
}
```

We've successfully retrieved the account from the blockchain, and it seems to contain the expected data, at least for the `topic` and `content` fields.

To finalize our test, let's incorporate assertions. For this purpose, we'll include the `assert` library at the beginning of our test file. No additional installations are necessary since it's already a part of our dependencies.

Now, instead of the previous console log, we'll employ assertions from the `assert` library to do the following:

- Verify equality between two entities via `assert.equal(actualThing, expectedThing)`
- Check if something is truthy with `assert.ok(something)`

Let's update our test accordingly, removing the console log and integrating these assertions:

```
Need to import: import { assert } from "chai";
it('can send a new message', async () => {
    const message = anchor.web3.Keypair.generate();
    await program.methods.sendMessage('space exploration','Discovering
new worlds!').accounts({message: message.publicKey})
    .signers([message]).rpc();
const messageAccount = await program.account.message.fetch(message.
publicKey);
```

```
    assert.equal(messageAccount.author.toBase58(), program.provider.
publicKey.toBase58());
    assert.equal(messageAccount.topic, 'space exploration');
    assert.equal(messageAccount.content, 'Discovering new worlds!');
    assert.ok(messageAccount.timestamp);
});
```

There are some things to note here:

- The initial assertion checks the match between the author of the new account and the public key of our wallet. It converts both `messageAccount.author` and `program.provider.wallet.publicKey` into `Base58` format using the `toBase58` method, ensuring equality if the resulting strings match. Solana wallet addresses use `Base58` encoding of public keys.

- Subsequent assertions validate the accurate storage of the message's topic and content.

- The final assertion confirms the presence of a non-empty timestamp within the tweet account. While verifying the timestamp's alignment with the current time is possible, it might lead to intermittent test failures due to precise time matching. Hence, the test focuses on ensuring the existence of a timestamp.

Everything's set! Running `anchor test` will confirm the success of our test and validate each of its assertions.

We have reached the end of the section, where we delved into testing our Solana program. From sending messages to the program to fetching newly created accounts and validating their content, we established a robust testing environment. By using assertions and confirming the accuracy of stored data, we've ensured the correctness and reliability of our Solana application. This process not only validates the functionality but also lays the groundwork for further development and integration with frontend applications.

Summary

In this chapter, we embarked on an exploration of Solana's development landscape, focusing on practical aspects of account structuring, sizing, and safeguarding against invalid data. We delved into nuances of Solana's data types, storage sizes, and the critical role of maintaining account structures for efficient program execution.

The journey began with an understanding of sizing guidelines for various data types in Solana, aiding developers in estimating storage sizes effectively. A detailed discussion provided insights into storage occupation, facilitating optimal account structuring. Moving forward, we examined the significance of storing the author's public key within the Solana program, elucidating its role in ensuring user-specific actions and maintaining data integrity. Safeguarding against invalid data was a key focus, wherein we implemented checks within the program to ensure data adherence to predefined constraints. These checks aimed to prevent scenarios of data overflow or invalid inputs. Transitioning to practical implementation, we crafted a Solana program for handling messages or tweets, meticulously defining account structures, sizing constraints, and integrating checks to validate inputs. The chapter underscored the strategic structuring of Solana programs, emphasizing the importance of sizing, validation, and efficient storage allocation.

As the chapter progressed, we shifted focus to writing tests for instructions, employing JavaScript clients to interact with the Solana blockchain. The testing phase revolved around executing transactions, validating account creations, and employing assertions to ensure data accuracy. The testing process concluded with successful validation, marking a crucial milestone in confirming the accuracy and reliability of our Solana program, setting the stage for further development and integration.

In the upcoming chapter, we will explore the NEAR blockchain, introducing its core components—accounts, transactions, and shards—to illuminate how they facilitate secure and efficient smart contract operations. We'll then dive into the intricacies of smart contract development on NEAR, covering everything from contract deployment and upgrades to testing and security best practices.

8

Exploring NEAR by Building a dApp

In the fast-paced world of blockchain technology, the NEAR Protocol shines as a pioneering platform for building **decentralized applications (dApps)**. Its remarkable focus on usability, scalability, and developer-friendly features has made it the preferred choice for diverse applications. This chapter provides an essential overview of NEAR and why it stands out in the world of dApp development.

The NEAR Protocol represents a paradigm shift in blockchain development. It offers a robust infrastructure with high performance and user-friendliness at its core. NEAR's unique sharded architecture ensures scalability by parallelizing transactions, enabling an impressive throughput of thousands of transactions per second. Its developer-centric approach abstracts complex blockchain mechanics, allowing developers to focus on creating secure and efficient smart contracts.

By the end of this chapter, you'll have a comprehensive understanding of NEAR's smart contract development, ready to leverage its transformative capabilities and contribute to the vibrant dApp ecosystem.

In this chapter, we'll cover the following key topics:

- Introducing NEAR
- Learning about the advanced concepts of NEAR
- Getting started with the NEAR blockchain
- Creating our first project with NEAR

Technical requirements

Here's what you will need to get started with the NEAR blockchain.

Prerequisites

To develop a smart contract, you will need to install Node.js. If you want to use Rust as your main language, then you'll need to install `rustup` as well.

You'll also need the following:

- **Node.js**: Download and install Node.js (`https://nodejs.org/en/download/`)
- **Rust and Wasm**:

  I.   Follow these instructions for setting up Rust: `https://doc.rust-lang.org/book/ch01-01-installation.html`.

  II.   Then, add the `wasm32-unknown-unknown` toolchain, which enables compiling Rust to WebAssembly (**Wasm**) – `https://webassembly.org/` – the low-level language used by the NEAR platform:

```
# Installing Rust in Linux and MacOS
curl --proto '=https' --tlsv1.2 https://sh.rustup.rs -sSf |
sh
source $HOME/.cargo/env

# Add the wasm toolchain
rustup target add wasm32-unknown-unknown
```

Next, we must set up the environment with the required installations.

Installation

To begin your journey with the NEAR Protocol and start developing dApps, you'll need to set up the necessary tools and environment. This installation guide will walk you through the process of installing the NEAR SDK and the NEAR **command-line interface** (**CLI**) on various platforms, including Ubuntu, Windows, and macOS. By following these steps, you'll be well-equipped to dive into the world of NEAR development.

Let's get started by ensuring you have all the prerequisites in place and have installed the essential components for NEAR development. Choose the instructions that match your operating system to seamlessly set up your development environment.

NEAR CLI installation

To install the NEAR CLI, you can follow these instructions for your preferred OS:

- **Ubuntu**:

  I.   Install Rust by running the following command:

  ```
  curl --proto '=https' --tlsv1.2 -sSf https://sh.rustup.rs |
  sh
  ```

  II. Add the Cargo `bin` directory to your PATH:

  ```
  source $HOME/.cargo/env
  ```

  III. Install the NEAR CLI by running the following command:

  ```
  cargo install near-cli --version 2.1.0
  ```

  IV. Verify the installation by running the following command:

  ```
  near --version
  ```

- **Windows**:

  I.   Ensure that you have WSL activated and openup your WSL terminal.

  II. Install Rust by running the following command:

  ```
  curl --proto '=https' --tlsv1.2 -sSf https://sh.rustup.rs |
  sh
  ```

  III. Add the Cargo `bin` directory to your PATH:

  ```
  echo 'export PATH="$HOME/.cargo/bin:$PATH"' >> ~/.bashrc
  source ~/.bashrc
  ```

  IV. Install the NEAR CLI by running the following command:

  ```
  cargo install near-cli --version 2.1.0
  ```

  V.   Verify the installation by running the following command:

  ```
  near -version
  ```

- **macOS**:

  I.   Install Homebrew by running the following command:

  ```
  /bin/bash -c "$(curl -fsSL https://raw.githubusercontent.
  com/Homebrew/install/HEAD/install.sh)"
  ```

II. Install Rust by running the following command:

```
curl --proto '=https' --tlsv1.2 -sSf https://sh.rustup.rs |
sh
```

III. Add the Cargo `bin` directory to your PATH:

```
echo 'export PATH="$HOME/.cargo/bin:$PATH"' >> ~/.zshrc
source ~/.zshrc
```

IV. Install the NEAR CLI by running the following command:

```
npm install -g near-cli
```

V. Verify the installation by running the following command:

```
near --version
```

Now that we have installed NEAR, let's talk about why the NEAR blockchain is a preferred choice to create dApps and explore the core concepts of NEAR.

Introducing NEAR

In this section, we'll embark on an exciting journey to explore the reasons why developers should choose the NEAR blockchain as their platform of choice for building dApps. We'll delve into the essential concepts that form the foundation of NEAR's ecosystem, starting with an in-depth understanding of accounts, addresses, and access keys. Furthermore, we'll unravel the intricacies of smart contracts and the blockchain state in NEAR, equipping developers with the knowledge to unlock the full potential of this innovative platform.

Why choose NEAR?

The NEAR blockchain offers compelling reasons for Rust developers to choose it as their preferred platform for developing and deploying smart contracts. Rust is a highly robust and secure programming language known for its memory safety and performance, and NEAR provides a seamless integration of Rust into its development stack, making it an ideal choice for Rust developers.

Here are the key factors that make NEAR an attractive option:

- **Native support for Rust**: NEAR natively supports Rust as one of its primary programming languages for smart contract development. This means that Rust developers can leverage their existing expertise and experience to build smart contracts on NEAR without the need to learn a new programming language or toolset. They can capitalize on Rust's safety features and its strong ecosystem of libraries and tools while benefiting from NEAR's unique capabilities.

- **Memory safety and performance**: Rust's memory safety features, such as ownership and borrowing, significantly reduce the risk of common programming errors such as null pointer dereferences and data races. This makes Rust an excellent choice for writing secure smart contracts. NEAR's integration with Rust ensures that developers can harness these advantages, resulting in highly reliable and performant smart contracts.

- **Scalability and high throughput**: NEAR's sharded architecture and advanced consensus mechanism enable high scalability and transaction throughput. Rust developers can build applications that can handle thousands of transactions per second, making NEAR well-suited for demanding use cases with a large user base or complex interactions.

- **Developer-friendly environment**: NEAR offers a developer-friendly environment with robust tooling, extensive documentation, and a supportive community. Developers can access a range of resources, including the NEAR SDKs, code examples, and tutorials, to accelerate their learning and development process. The NEAR ecosystem fosters collaboration and knowledge sharing, allowing developers to engage with like-minded individuals and benefit from shared experiences.

- **Ecosystem opportunities**: NEAR has been gaining traction in the blockchain space, attracting numerous projects and collaborations. By choosing NEAR, Rust developers can tap into a thriving ecosystem of dApps, tools, and services. This presents opportunities for collaboration, partnerships, and building decentralized solutions that can reach a broader user base.

NEAR's native support for Rust, combined with its scalability, performance, developer-friendly environment, and growing ecosystem, positions it as an excellent choice for Rust developers seeking to develop and deploy smart contracts. By leveraging their Rust expertise, developers can build secure, efficient, and scalable dApps on the NEAR blockchain.

Understanding the foundational elements of NEAR

Now, let's explore the foundational elements of NEAR's architecture, which are accounts, addresses, access keys, smart contracts, and state.

Accounts

In the NEAR ecosystem, **accounts** play a pivotal role, serving as the entry point for users to interact with dApps. Understanding the intricacies of accounts is crucial for developers seeking to build secure and user-friendly applications on the NEAR blockchain. Let's dive deep into the details of account IDs, implicit accounts, and named accounts, providing insightful code examples along the way:

- **Account ID**: At the core of NEAR's account structure is the account ID. It is a unique identifier associated with each account on the NEAR blockchain. Account IDs are human-readable and are typically represented as strings. For instance, `alice.near` or `my_dapp.near` could be valid account IDs. Account IDs act as the addresses to which funds can be sent and provide access to the associated account's data and smart contract functionality.

- **Implicit accounts**: Implicit accounts are a special type of account in NEAR that are created automatically as part of a transaction. When a transaction is sent from a particular account ID that does not exist, NEAR automatically creates an implicit account with that ID. Implicit accounts are useful for one-time interactions or temporary data storage within a transaction. They don't have associated private keys and cannot receive funds or store long-term data. Here's an example of implicit account creation in Rust:

```
// Create an implicit account in a transaction
#[near_bindgen]
pub fn create_implicit_account(&mut self, account_id: String) {
    let account_id: ValidAccountId = account_id.try_into().
unwrap();
    env::log(format!("Creating implicit account: {}", account_
id).as_bytes());

    // Perform actions with the implicit account
    // ...
}
```

- **Named accounts**: Named accounts are accounts with persistent state and private keys. They can receive funds, store data, and interact with other smart contracts. Named accounts provide a more permanent identity for users or dApps within the NEAR ecosystem. Developers can create and manage named accounts programmatically using NEAR's SDKs.

Understanding the distinctions between account IDs, implicit accounts, and named accounts provides developers with the flexibility to design tailored solutions for various use cases within the NEAR ecosystem. By utilizing the power of accounts, developers can enable secure and seamless interactions for their users and build robust dApps on the NEAR blockchain.

Now, we will delve into the details of NEAR addresses, exploring their structure and their functions within the NEAR blockchain ecosystem.

Addresses

Addresses are a vital component of the NEAR blockchain, serving as unique identifiers for accounts and facilitating secure transactions and interactions within the network. Understanding addresses is essential for developers seeking to build dApps on the NEAR blockchain. Let's delve into the intricacies of addresses, shedding light on their structure and significance.

In the NEAR ecosystem, addresses are derived from account IDs, which act as human-readable representations of accounts. NEAR addresses are cryptographic hashes generated from account IDs, providing a secure and tamper-resistant way to identify accounts. The cryptographic nature of addresses ensures the integrity and authenticity of transactions and interactions within the NEAR blockchain.

NEAR addresses have two primary functions:

- **Transaction routing**: NEAR addresses play a crucial role in routing transactions to the intended recipients. When initiating a transaction, the sender specifies the destination account's address to ensure that the transaction reaches the correct recipient. The NEAR network utilizes addresses to determine the appropriate shard, or a subset of nodes, responsible for processing the transaction, enabling efficient and scalable transaction processing.

- **Secure transactions and interactions**: NEAR addresses provide a cryptographic layer of security. The private key associated with an address is used to sign transactions, verifying the authenticity and integrity of the sender. This ensures that only the account owner or authorized entity can initiate transactions on behalf of that account. NEAR's address structure and cryptographic mechanisms enhance the security of funds, data, and smart contract interactions within the ecosystem.

Developers working with NEAR can leverage addresses to enable secure and seamless user experiences. By incorporating addresses into their dApps, developers can enable users to interact with the blockchain securely, send and receive funds, and access specific functionalities within their applications.

Now, let's talk about access keys and their pivotal role within the NEAR Protocol, encompassing account management, security, and enhancing flexibility throughout the ecosystem.

Access keys

Access keys are a fundamental component of the NEAR Protocol, playing a vital role in account management, security, and flexibility within the ecosystem.

Access keys provide a powerful mechanism for managing account permissions and controlling access to accounts and their associated resources within the NEAR Protocol. They offer several key benefits:

- **Enhanced security**: Access keys allow fine-grained control over the capabilities granted to different entities interacting with an account. This enables users to limit access to third-party applications or smart contracts, reducing the risk of unauthorized actions.

- **Granular permissions**: Access keys enable developers to grant specific permissions to different entities. This allows for modular and secure account management, ensuring that only authorized operations can be performed by designated entities.

- **Flexibility and delegation**: Access keys provide the flexibility to delegate certain functionalities to other accounts or smart contracts, enabling seamless integration and interaction between dApps. This facilitates complex multi-signature setups and delegation of authority for account management.

The NEAR Protocol supports two types of access keys:

- **Full access keys**: Full access keys grant complete control over an account, allowing the holder to perform any operation on behalf of the account. These keys are typically used by account owners or trusted entities requiring full control over the associated account.

- **Function call keys**: Function call keys, also known as limited access keys, grant permissions for specific actions or function calls within a smart contract. They allow fine-grained control over what operations can be performed on the account. This key type is commonly used for the delegation of specific tasks to trusted third-party contracts or for executing specific actions without granting full account access.

Access keys are particularly important in the context of **locked accounts**, where the account owner designates a specific access key to manage the account's resources. A locked account requires the usage of a specific access key for any transaction or operation to be executed successfully. This provides an additional layer of security because even if an attacker gains access to other access keys associated with the account, they cannot perform any operation without the designated access key.

Here's an example of creating a full access key and a function call key in Rust:

```
// Create a full access key
#[near_bindgen]
pub fn create_full_access_key(&mut self, public_key: PublicKey) {
    self.env().key_create(
        public_key,
        &access_key::AccessKey {
            nonce: 0,
            permission: access_key::Permission::FullAccess,
        },
    );
}

// Create a function call key
#[near_bindgen]
pub fn create_function_call_key(&mut self, public_key: PublicKey) {
    self.env().key_create(
        public_key,
        &access_key::AccessKey {
            nonce: 0,
            permission: access_key::Permission::FunctionCall {
                allowance: access_key::FunctionCallPermission {
                    allowance: 10.into(),   // Maximum number of
function call allowances
                    receiver_id: "receiver_account".to_string(),
                    method_names: vec!["allowed_method".to_string()],
```

```
                },
            },
        },
    );
}
```

Access keys empower NEAR account holders to delegate permissions, enhance security, and enable flexible account management. By leveraging different key types and implementing access key strategies effectively, developers can build secure and dynamic dApps within the NEAR Protocol.

Now that we've learned about access keys and how they empower NEAR account holders, let's shift our focus to smart contracts, the foundational building blocks of dApps in the NEAR blockchain.

Smart contracts

Smart contracts form the backbone of dApps in the NEAR blockchain, enabling trustless interactions and automating business logic. In this section, we'll delve into the concept of smart contracts and their significance within the NEAR ecosystem. We'll also provide code examples to illustrate their implementation.

In smart contracts, mutability is a key feature that dictates whether a contract can be updated post-deployment. **Immutable contracts** offer high security and trust by locking the code against changes, ensuring the original rules remain intact but at the cost of being unable to correct any flaws. Conversely, **upgradeable contracts** allow for modifications to address bugs, enhance functionality, or meet evolving needs, offering flexibility but adding complexity and raising security concerns to ensure updates are securely managed and do not undermine the contract's or users' interests. NEAR addresses these challenges by providing developers with tools and governance models to strike a balance between security and adaptability, guiding them to choose the level of mutability that best suits their application's requirements and security needs.

Smart contracts are self-executing agreements that contain the rules and logic governing the interactions between participants in a decentralized system. In the NEAR blockchain, smart contracts are written using languages such as Rust, AssemblyScript, or TypeScript, and they reside on the blockchain, ensuring transparency and immutability.

Here's an example of a simple smart contract written in Rust using the NEAR SDK:

```
use near_sdk::borsh::{self, BorshDeserialize, BorshSerialize};
use std::collections::HashMap;
use near_sdk::near_bindgen;
use near_sdk::borsh;
#[near_bindgen]
#[derive(Default, borsh::BorshSerialize, borsh::BorshDeserialize)]
pub struct MyContract {
    data: HashMap<u64, u64>
```

```
}
#[near_bindgen]
impl MyContract {
    pub fn insert_data(&mut self, key: u64, value: u64) -> Option<u64>
{
        self.data.insert(key, value)
    }
    pub fn get_data(&self, key: u64) -> Option<u64> {
        self.data.get(&key).cloned()
    }
}
```

In this example, we define a smart contract called MyContract.

Smart contracts in the NEAR blockchain offer several advantages. They provide transparency and immutability, ensuring that contract logic and data cannot be tampered with once deployed. The NEAR blockchain's execution environment guarantees security and fault tolerance, ensuring that contracts operate as intended. Additionally, smart contracts enable the automation of business logic, reducing the need for intermediaries and increasing efficiency.

Next, we will explore the concept of state in the NEAR ecosystem, focusing on account metadata and contract state since they play crucial roles in managing data within the blockchain.

State management

State management facilitates the storage and retrieval of data associated with accounts and smart contracts. Let's explore the concept of state in the NEAR ecosystem with a focus on **account metadata** and **contract state**.

In the NEAR blockchain, accounts can store metadata alongside their basic information. Account metadata includes additional details such as profile information, user preferences, or any other custom data relevant to the account. This metadata provides a flexible way to personalize and enhance the user experience within dApps. Developers can store and update account metadata using the NEAR SDKs and access it within their applications.

Here's an example of storing account metadata:

```
// Import necessary libraries
use near_sdk::borsh::{self, BorshDeserialize, BorshSerialize};
use near_sdk::env;
use near_sdk::near_bindgen;

// Define the contract structure
#[near_bindgen]
#[derive(Default, BorshDeserialize, BorshSerialize)]
```

```
pub struct YourContract {
    // Declare a field to store account metadata
    pub account_metadata: Option<String>,
}

// Implement methods for storing and retrieving account metadata
#[near_bindgen]
impl YourContract {
    // Method to set or update account metadata
    pub fn set_account_metadata(&mut self, metadata: String) {
        self.account_metadata = Some(metadata);
    }

    // Method to retrieve account metadata
    pub fn get_account_metadata(&self) -> Option<String> {
        self.account_metadata.clone()
    }
}
```

Smart contracts on the NEAR blockchain maintain their state, which represents the data and variables associated with the contract's logic and functionality. The contract state includes any information required for the contract to perform its operations and maintain its internal state. This can include user balances, storage of user-specific data, and any other relevant contract-specific information. The contract state is persisted on the blockchain and can be accessed and modified through contract methods.

Here's an example of a smart contract in Rust on the NEAR blockchain, showcasing the management of a contract state:

```
use near_sdk::borsh::{self, BorshDeserialize, BorshSerialize};
use near_sdk::near_bindgen;

#[near_bindgen]
#[derive(Default, BorshDeserialize, BorshSerialize)]
pub struct MyContract {
    counter: i32,
}

#[near_bindgen]
impl MyContract {
    pub fn new() -> Self {
        Self { counter: 0 }
    }

    pub fn increment(&mut self) {
```

```
        self.counter += 1;
    }

    pub fn get_counter(&self) -> i32 {
        self.counter
    }
}
```

In this example, the MyContract struct represents the smart contract with the counter variable as its state. The new function initializes the state with a default value of 0. The increment function increments the counter, and the get_counter function retrieves its value.

This Rust implementation utilizes the NEAR SDK's attributes and derives serialization traits using the borsh crate. These attributes and traits ensure that the contract state is properly serialized/deserialized when interacting with the NEAR blockchain.

Now that we have learned about the core concepts of the NEAR blockchain, it's time to dive into some advanced concepts that will be helpful when we start building out our projects on NEAR.

Learning about the advanced concepts of NEAR

In this section, we'll delve into the advanced concepts of the NEAR blockchain, expanding our understanding of its inner workings and exploring key topics that enable developers to build robust and scalable dApps. We will cover essential concepts such as **transactions** and **gas**, **data flow**, **token management**, **storage optimization**, **validator networks**, and the NEAR SDK.

First, we'll talk about transactions and gas.

Transactions and gas

Transactions are the fundamental units of activity in the NEAR blockchain. They represent actions performed by accounts, such as transferring tokens, calling contract methods, or updating account data. To execute a transaction, users need to allocate gas, a unit of computation and storage resource, which determines the complexity and cost of the transaction.

Let's explore a code example of a simple transaction using the NEAR SDK for Rust:

```
use near_sdk::borsh::{self, BorshDeserialize, BorshSerialize};
use near_sdk::env;
use near_sdk::near_bindgen;
use near_sdk::serde_json::*;
#[near_bindgen]
#[derive(Default, BorshDeserialize, BorshSerialize)]
pub struct MyContract {
    // Contract state and functionality
```

```
}
#[near_bindgen]
impl MyContract {
    pub fn transfer_tokens(&mut self, receiver_id: String, amount:
u64) {
        let current_account_id = env::current_account_id();
        let transfer_action = json!({ "receiver_id": receiver_id,
"amount": amount });
        let transfer_call = json!({ "contract": current_account_id,
"method": "transfer", "args": transfer_action, "gas": env::prepaid_
gas() - near_sdk::Gas(100_000_000), // Subtracting 100 TeraGas for
additional actions
 "attached_gas": near_sdk::Gas(5_000_000_000), // 5 GigaGas attached
to cover the cost
  "attached_tokens": amount });
        env::promise_create(
            env::current_account_id(),
            "do_transfer",
            transfer_call.to_string().as_bytes(),
            amount.into(),
            0.into(),
        );
    }
}
```

In this code snippet, we define a MyContract struct with a transfer_tokens method that performs a token transfer to a specified receiver account. The gas allocation is crucial for executing the transaction successfully. Here, we subtract a certain amount of prepaid gas to accommodate additional actions within the same transaction.

Gas allocation helps ensure that the NEAR blockchain's resources are fairly and efficiently utilized. It incentivizes users to optimize their contracts and limit resource-intensive operations to minimize costs. By managing gas allocation effectively, developers can design efficient and cost-effective transactions, promoting scalability and user-friendly experiences within their dApps on the NEAR blockchain.

Data flow

Understanding the data flow within the NEAR blockchain is essential for designing efficient and secure dApps. We will examine how data is propagated, processed, and stored across the network, and explore strategies for optimizing data management and access. The data flow within the NEAR blockchain may initially appear complex, but it follows well-defined rules and can be easily understood. In this section, we will explore the intricacies of data flow within the NEAR blockchain in greater detail.

To comprehend the data flow, it is helpful to visualize it as an infinite timeline with a starting point but no endpoint. Blocks appear at regular intervals on this timeline, and each block contains information about the previous block, forming a chain of interconnected blocks. The following diagram helps us visualize this:

Figure 8.1 – Diagram to visualize the data flow

The NEAR Protocol employs a sharded nature, which means that multiple parallel networks, known as **shards**, can be active simultaneously. Each shard generates a chunk of a block at specific intervals. In the NEAR blockchain, a block is a collection of these block chunks from all shards. In the NEAR Protocol documentation, a block chunk is referred to as a **chunk**.

To better understand the data flow, we can imagine **tracks** similar to those found in audio/video editing applications. Each shard has its own set of tracks, and the top track is reserved for chunks. Chunks appear at fixed intervals, typically around one second in the NEAR blockchain. It's important to note that chunks continue to be produced even if no blockchain activity is occurring.

By visualizing the data flow within the NEAR blockchain as tracks and chunks, we can gain a clearer understanding of how information is organized and processed. This understanding is essential for developers and users alike to navigate the NEAR ecosystem effectively and leverage its capabilities to build and interact with dApps.

The following diagram helps us visualize the data flow as tracks and chunks:

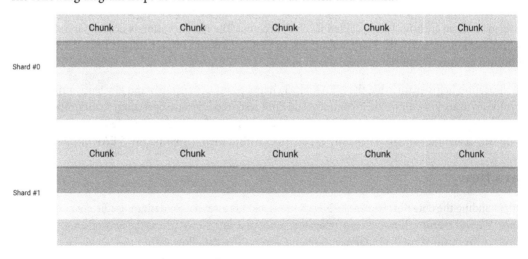

Figure 8.2 – NEAR data flow – chunks and shards

When we say that something is happening within the NEAR blockchain, we refer to the occurrence of events that trigger changes within the blockchain. The primary method to initiate these changes is by sending a transaction to the blockchain, which contains instructions specifying the desired modifications and the identity of the requester.

To execute a transaction, it must be meticulously constructed, signed, and subsequently transmitted to the blockchain. After execution, we anticipate receiving a result known as the ExecutionOutcome. However, it is important to note that the simplicity of this concept is not entirely accurate when applied to the NEAR blockchain.

The NEAR blockchain introduces a unique twist to the transaction process. Rather than immediately providing a definitive outcome, transactions within the NEAR blockchain have a conditional nature. This means that `ExecutionOutcome` may vary, depending on the state of the blockchain during the transaction's execution.

Factors such as network congestion, concurrent transactions, and conflicting state changes can influence the outcome of a transaction. As a result, developers and users need to account for these conditional outcomes when designing and interacting with smart contracts on the NEAR blockchain.

The following diagram illustrates how transactions and receipts can be visualized concerning the chunks and tracks as part of the data flow:

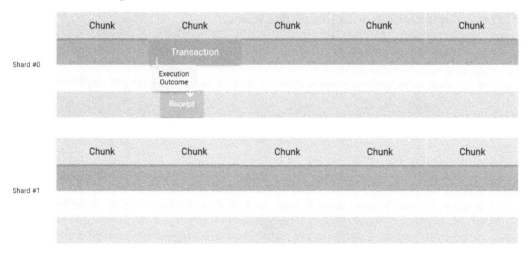

Figure 8.3 – Token transfer flow on NEAR

At the inception of a transaction in the NEAR blockchain, it contains the instructions that we wish to execute on the blockchain. This transaction is then sent to the NEAR blockchain for processing.

However, it's important to note that the immediate outcome of the transaction execution is merely an acknowledgment indicating that the transaction will be executed on the blockchain. This internal execution request is known as a **receipt**. Conceptually, you can envision the receipt as an internal transaction that facilitates the transfer of information across different shards within the NEAR blockchain.

Let's revisit the analogy of tracks to illustrate this process. Suppose we have two accounts, `alice.near` and `bob.near`, residing on different shards. Alice initiates a transaction to transfer tokens to `bob.near`. Upon execution, the transaction generates `ExecutionOutcome`, which is always in the form of a receipt.

However, this receipt cannot be directly executed on the current shard because `bob.near` resides on a different shard. Consequently, the receipt must be transferred to the shard where `bob.near` is located.

Once the receipt reaches the destination shard, it is executed, and the process is considered complete. At this point, the tokens have been successfully transferred from `alice.near` to `bob.near`.

Thus, the overall scheme can be visualized as a transaction originating from one account, moving through the blockchain's internal processes, and ultimately, being executed on the appropriate shard where the recipient account resides. This ensures that transactions and associated receipts are processed accurately and efficiently within the NEAR blockchain.

The following diagram helps us to visualize the final schema:

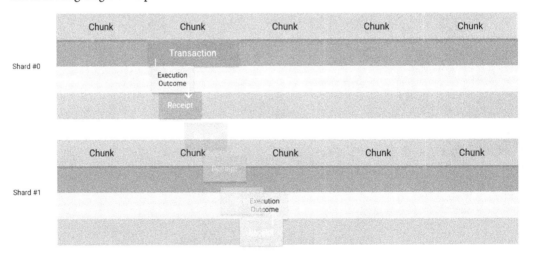

Figure 8.4 – Token transfer flow on NEAR – sharded execution

Now, let's delve into tokens and practical strategies for safeguarding tokens within the NEAR blockchain ecosystem.

Tokens and avoiding loss

Tokens play a vital role in the NEAR blockchain ecosystem, serving as a means of value transfer and enabling various functionalities within dApps. However, it is crucial to implement measures to prevent token loss, for reasons such as loss of access key and refunding deleted accounts (when a refund receipt is issued for an account, if that account no longer exists, the funds will be dispersed among validators proportional to their stake in the current epoch), and to ensure secure token management. Here are two ways developers can avoid token loss in the NEAR blockchain, demonstrated using code examples:

- **Implementing token transfer safeguards**: When transferring tokens, it is essential to include checks and validations to prevent accidental loss. For instance, developers can verify that the recipient account exists and is valid before initiating the transfer. Here's an example in Rust using the NEAR SDK:

```
pub fn transfer_tokens2(self, recipient: near_sdk::AccountId,
amount: near_sdk::Balance) {
        assert!(
            env::is_valid_account_id(&recipient.as_bytes()),
            "Invalid recipient account"
        );
        let sender_balance = account_balance();
        assert!(sender_balance >= amount, "Insufficient
balance"); // Perform the token transfer
        Promise::new(recipient).transfer(amount);
    }
```

- **Implementing token vesting**: Token vesting is a mechanism that gradually releases tokens over a specified period to prevent immediate loss or misuse. Developers can create smart contracts with token vesting functionality to lock tokens and release them gradually based on predetermined conditions. Here's an example using the NEAR SDK:

```
pub struct TokenVesting {
    beneficiary: AccountId,
    start_timestamp: u64,
    duration: u64,
    total_tokens: Balance,
}

impl TokenVesting
        pub fn release_tokens(&mut self) {
        let current_timestamp = env::block_timestamp();
        let elapsed_time: u128 = (current_timestamp - self.
start_timestamp).into();
        let duration: u128 = self.duration.into();
        let beneficiary = self.beneficiary.clone();
```

```
        if elapsed_time >= duration {
            Promise::new(beneficiary).transfer(self.total_
tokens);
        } else {
            let tokens_to_release = (self.total_tokens *
elapsed_time) / duration;
            Promise::new(beneficiary).transfer(tokens_to_
release);
        }
    }
}
```

By implementing safeguards in token transfers and incorporating token vesting mechanisms, developers can mitigate the risk of token loss and enhance the security of token management within the NEAR blockchain. These practices promote the responsible handling of tokens and contribute to the overall integrity of dApps.

Now, let's explore the various storage options that NEAR provides to developers for efficient and organized data storage within smart contracts. We'll also provide code examples to illustrate their usage.

Storage options

NEAR provides developers with various data structures for efficient and organized storage of data within smart contracts. Let's explore some of the storage options available on the NEAR blockchain along with code examples:

- **Vector**: Vectors are dynamic arrays that allow efficient storage and retrieval of elements. They are useful for collections that can grow or shrink over time. In the following code example, we declare a vector of unsigned 64-bit integers ($u64$) and demonstrate how to add elements to it and access specific elements:

  ```
  // Declare a vector of u64 elements
  let mut my_vector: Vec<u64> = Vec::new();

  // Add elements to the vector
  my_vector.push(10);
  my_vector.push(20);
  my_vector.push(30);

  // Access elements in the vector
  let second_element = my_vector[1];
  ```

- **LookupSet**: LookupSet is an unordered collection of unique elements. It provides efficient membership checks. In the following code example, we declare LookupSet of `string` elements, insert values into it, and check for membership:

```
// Declare a LookupSet of string elements
let mut my_lookupset: LookupSet<String> = LookupSet::new();

// Add elements to the LookupSet
my_lookupset.insert("apple".to_string());
my_lookupset.insert("banana".to_string());

// Check membership
let contains_apple = my_lookupset.contains("apple".to_string());
```

- **UnorderedSet**: UnorderedSet is a collection of unique elements. It enables quick iteration over stored elements. In the following code example, we declare `UnorderedSet` of unsigned 32-bit integers (u32), add elements to it, and demonstrate how to iterate over the elements:

```
// Declare an UnorderedSet of u32 elements
let mut my_unorderedset: UnorderedSet<u32> =
UnorderedSet::new();

// Add elements to the UnorderedSet
my_unorderedset.insert(1);
my_unorderedset.insert(2);

// Iterate over the elements
for element in my_unorderedset.iter() {
    // Process each element
}
```

- **LookupMap**: LookupMap is a key-value data structure that provides efficient value retrieval based on a given key. In the following code example, we declare a LookupMap structure with `string` keys and unsigned 64-bit integer values (u64), insert key-value pairs, and access values based on keys:

```
// Declare a LookupMap with string keys and u64 values
let mut my_lookupmap: LookupMap<String, u64> = LookupMap::new();

// Add key-value pairs to the LookupMap
my_lookupmap.insert("key1".to_string(), 10);
my_lookupmap.insert("key2".to_string(), 20);

// Access values based on keys
let value = my_lookupmap.get("key1".to_string());
```

- **UnorderedMap**: UnorderedMap is a key-value data structure that does not guarantee any specific order of elements. In the following code example, we declare an UnorderedMap structure with unsigned 32-bit integer keys (u32) and string values, insert key-value pairs, and demonstrate how to iterate over the key-value pairs:

```
// Declare an UnorderedMap with u32 keys and string values
let mut my_unorderedmap: UnorderedMap<u32, String> =
UnorderedMap::new();

// Add key-value pairs to the UnorderedMap
my_unorderedmap.insert(1, "value1".to_string());
my_unorderedmap.insert(2, "value2".to_string());

// Iterate over the key-value pairs
for (key, value) in my_unorderedmap.iter() {
    // Process each key-value pair
}
```

- **TreeMap**: TreeMap is a data structure that maintains its elements in sorted order based on the keys. In the following code example, we declare a TreeMap structure with unsigned 64-bit integer keys (u64) and string values, insert key-value pairs, and demonstrate how to iterate over the key-value pairs in sorted order:

```
// Declare a TreeMap with u64 keys and string values
let mut my_treemap: TreeMap<u64, String> = TreeMap::new();

// Add key-value pairs to the TreeMap
my_treemap.insert(3, "value3".to_string());
my_treemap.insert(1, "value1".to_string());
my_treemap.insert(2, "value2".to_string());

// Iterate over the key-value pairs in sorted order
for (key, value) in my_treemap.iter() {
    // Process each key-value pair
}
```

Now, let's dive into the vital role that validators play in the NEAR network and explore the consensus mechanism known as **Proof of Stake (PoS)** that's employed by NEAR. We'll also discuss how validators contribute to the network's integrity and governance.

Validators and consensus

Validators are responsible for validating transactions and blocks in the NEAR network. They maintain the network's decentralized nature by participating in the consensus process. Validators are chosen based on their stake and reputation, and they actively contribute to block production and network governance.

NEAR employs a consensus mechanism called PoS. In this mechanism, validators are selected based on the number of tokens they hold and are willing to **stake** as collateral. Validators who stake a significant amount of tokens have a higher chance of being selected to produce blocks and participate in consensus.

The consensus process in NEAR involves validators collectively agreeing on the state of the blockchain and validating transactions. They reach a consensus on the order of transactions and the validity of blocks, ensuring that the network remains secure and free from double-spending or malicious activities.

Validators in the NEAR blockchain earn rewards for their participation and successful block production. These rewards incentivize validators to act honestly and maintain the network's integrity. They are also responsible for validating and executing smart contracts, ensuring the accuracy and consistency of the blockchain state.

Through the validators and consensus mechanism, the NEAR blockchain achieves distributed consensus, enabling a secure and decentralized environment for developers to build and deploy applications. Validators contribute to the network's governance, maintain its security, and enable the smooth functioning of the NEAR ecosystem, fostering trust and reliability for all participants.

Next, we'll talk about the NEAR SDK, a powerful toolkit tailored to streamline the development of dApps on the NEAR blockchain.

NEAR SDK

In multiple sections of this chapter, we have briefly mentioned the NEAR SDK, but now, it's time to understand it in more detail.

The NEAR SDK is a powerful toolset that empowers developers to build dApps on the NEAR blockchain. The NEAR SDK provides a collection of libraries, APIs, and tools that simplify the development process, enabling developers to interact with the NEAR blockchain and build innovative applications.

Recently, certain functionalities of the NEAR SDK have been reorganized. Some of the components that were part of the NEAR SDK are now available in a separate crate called `near_cli`. This crate is specifically designed to offer command-line tools and utilities for interacting with the NEAR blockchain.

To install `near_cli`, you can use the following command:

```
cargo install near-cli-rs
```

For more detailed information and the latest updates on `near_cli`, please refer to the official documentation at `https://docs.near.org/tools/near-cli-rs`.

Let's explore some key features of the NEAR SDK, along with the functionalities provided by near_cli, through code examples:

- **Account creation and contract deployment**: The NEAR SDK allows developers to create accounts and deploy smart contracts seamlessly. Here's an example:

```
const NEAR_RPC_URL: &str = "https://rpc.mainnet.near.org";

// Connect to the NEAR network
let near = near_sdk::connect::connect(near_sdk::Config {
    network_id: "mainnet".to_string(),
    node_url: NEAR_RPC_URL.to_string(),
});

// Create a new account using async/await and Promises API
async fn create_and_deploy() {
    let new_account = near.create_account("new_account").await.
unwrap();

    // Load contract code
    let contract_code = include_bytes!("path/to/contract.wasm");

    // Deploy a contract to the new account using Promises API
    new_account.deploy_contract(contract_code).await.unwrap();
}

// You can call `create_and_deploy` function in an async context
```

- **Interacting with smart contracts**: The NEAR SDK provides APIs for interacting with smart contracts, making it easy to call contract methods and retrieve data. Here's an example:

```
// Instantiate a contract object
let contract = Contract::new(account_id, contract_id, signer);

// Call a method on the contract
contract.call_method("method_name", json!({ "param": "value"
}));

// Get contract state
let state: ContractState = contract.view_method("get_state",
json!({}));
```

- **Handling tokens**: The NEAR SDK offers functionalities to handle NEAR tokens within your dApps. Developers can transfer tokens, check balances, and implement token-related logic. Here's an example:

```
// Transfer tokens from one account to another
let sender = near.get_account("sender_account");
let recipient = near.get_account("recipient_account");
sender.transfer(&recipient, 100);

// Check token balance
let balance = recipient.get_balance();
```

The NEAR SDK provides a comprehensive set of tools and APIs that simplify the development process for building dApps on the NEAR blockchain. Its intuitive interface, smart contract deployment capabilities, and token-handling features make it a valuable resource for developers seeking to leverage the NEAR ecosystem and build innovative dApps.

Now, let's build our very first NEAR blockchain.

Getting started with the NEAR blockchain

At this point, we have clarity on most of the important concepts of NEAR Protocol. Now, we'll get a bit more hands-on and delve into the core concepts of developing the NEAR blockchain. In this section, we will cover essential topics and tools that will equip you with the necessary knowledge to build robust dApps on NEAR.

By the end of this section, you will have a solid understanding of the core concepts and tools necessary for developing sophisticated dApps on the NEAR blockchain, empowering you to create innovative solutions within this thriving ecosystem.

First, we'll explore the `Contract` class, which serves as the foundation for your smart contracts on NEAR. We will delve into the various modules, types, and structs that can be used to define and manipulate data within your contracts.

The Contract class

In NEAR Protocol, the `Contract` class provides a set of functionalities that allow developers to define and interact with smart contracts. The `Contract` class is written in Rust and serves as the bridge between the blockchain and the dApp.

Here's an example of how to define and use the `Contract` class in Rust:

```
use near_sdk::borsh::{self, BorshDeserialize, BorshSerialize};
use near_sdk::collections::Vector;
use near_sdk::{env, near_bindgen, AccountId, Balance, PanicOnDefault,
Promise, StorageUsage};
```

Let us now complete the Contract class:

1. We start by importing the necessary dependencies from the NEAR SDK:

```
#[near_bindgen]
#[derive(BorshDeserialize, BorshSerialize, PanicOnDefault)]
pub struct MyContract {
    pub items: Vector<String>,
}
```

2. Next, we define the main contract struct, MyContract, with its associated methods:

```
impl MyContract {
    pub fn new() -> Self {
        Self {
            items: Vector::new(b"i".to_vec()),
        }
    }
}
```

3. The new() constructor function initializes the contract, including the vector of items:

```
    pub fn add_item(&mut self, item: String) {
        self.items.push(&item);
    }
```

4. The add_item() function allows us to add items to the vector:

```
    pub fn get_items(&self) -> Vec<String> {
        self.items.to_vec()
    }
```

5. The get_items() function retrieves all the items from the vector:

```
#[near_bindgen]
impl MyContract {
    pub fn contract_metadata(&self) -> ContractMetadata {
        ContractMetadata {
            name: "MyContract".to_string(),
            version: "1.0.0".to_string(),
            // Additional metadata fields...
        }
    }
}
```

6. We define additional `contract` functions, including `contract_metadata()`, to provide metadata:

```
#[derive(Default, BorshDeserialize, BorshSerialize)]
pub struct ContractMetadata {
    pub name: String,
    pub version: String,
    // Additional metadata fields...
}
```

7. Finally, we define the `ContractMetadata` struct to store contract-related information.

In this example, we define a `MyContract` struct with a `Vector` collection to store items. The `add_item` and `get_items` functions modify and retrieve the items, respectively. The `contract_metadata` function returns metadata about the contract.

The `Contract` class simplifies the development process by providing an intuitive interface for interacting with smart contracts on the NEAR blockchain. It encapsulates the necessary logic for serialization, deserialization, and interaction with blockchain data structures, allowing developers to focus on building the core functionality of their dApps.

Now, let's look into the serialization protocols of NEAR.

The serialization protocols of NEAR

Serialization in the context of the NEAR blockchain is a critical process that transforms data structures or object states into a format that can be stored or transmitted and subsequently reconstructed. NEAR uses efficient **serialization protocols** to ensure that data exchanged between smart contracts and the blockchain is compact, fast to process, and secure. These protocols play a vital role in optimizing the performance and interoperability of dApps on the NEAR platform.

The primary serialization protocol used by NEAR is **Borsh**, which stands for **Binary Object Representation Serializer for Hashing**. Borsh is designed for optimal speed and safety, aiming to be both compact and fast for reading and writing operations. It ensures deterministic serialization, which is crucial for blockchain transactions where consistency and predictability of data representation are necessary. By leveraging Borsh, developers can ensure their smart contracts are efficient in terms of storage and speed, thereby reducing execution costs and enhancing user experience.

Borsh's deterministic nature also contributes to the security and integrity of smart contracts by preventing ambiguities in data representation that could lead to vulnerabilities. This serialization protocol supports various data types, including primitives, structs, enums, and collections, making it a versatile choice for NEAR developers.

For those interested in diving deeper into the specifics of serialization, handling contract calls, and managing data within the NEAR blockchain, additional information and technical details are available. You can explore further explanations and examples by visiting this link: `https://docs.near.org/sdk/rust/contract-interface/serialization-interface`.

We'll now shed light on state and data structures in the NEAR Protocol's smart contract landscape.

State and data structures

In the NEAR Protocol, **state** and **data structures** form the backbone of smart contracts, enabling the storage and manipulation of information on the blockchain. Let's explore these concepts further:

- **State** refers to the persistent data stored on the blockchain that represents the current state of a smart contract. It captures the contract's data, variables, and values that can be accessed and modified by contract functions. NEAR employs a key-value store model for storing and retrieving the contract's state, making it efficient and accessible. State changes are transactional, ensuring atomicity and consistency.

- **Data structures** define how the state is organized and stored within a smart contract. NEAR provides various built-in data structures, such as maps, vectors, and sets, to store and manipulate data efficiently. Developers can also create custom data structures using structs and enums, enabling them to model complex data relationships.

For example, consider a smart contract that manages a list of users' information. It could use a NEAR map data structure to store user details, such as `name`, `age`, and `address`, with their `account ID` values serving as keys:

```rust
use near_sdk::borsh::{self, BorshDeserialize, BorshSerialize};
use near_sdk::collections::LookupMap;
use near_sdk::{env, near_bindgen, AccountId};
```

Let's now begin creating this smart contract:

1. Import the necessary dependencies and modules from the NEAR SDK:

   ```rust
   #[near_bindgen]
   #[derive(BorshDeserialize, BorshSerialize)]
   pub struct UserRegistry {
       users: LookupMap<AccountId, UserInfo>,
   }
   ```

2. Define the `UserRegistry` smart contract struct, which includes a `users` field of the Map type to store user information:

   ```rust
   #[derive(BorshDeserialize, BorshSerialize)]
   pub struct UserInfo {
       name: String,
       age: u32,
       address: String,
   }
   ```

3. Define the `UserRegistry` struct, which represents the data structure for user information, including name, age, and address:

```
impl UserRegistry {
    pub fn new_user(&mut self, name: String, age: u32, address:
String) {
        let caller = env::signer_account_id();
        let user_info = UserInfo { name, age, address };
        self.users.insert(&caller, &user_info);
    }
    pub fn get_user(&self, user_id: AccountId) ->
Option<UserInfo> {
        self.users.get(&user_id)
    }
}
```

4. Implement methods for the `UserRegistry` smart contract:

    *   `new_user`: This creates a new user and stores their information in the `users` map. It takes parameters for `name`, `age`, and `address` and associates the user data with the caller's account ID.

    *   `get_user`: This retrieves user information based on their account ID from the `users` map and returns it as `Option<UserInfo>`.

In this example, `UserRegistry` uses a NEAR map (`users`) to store user information using account IDs as keys. The `new_user` function adds a new user to the registry, and `get_user` retrieves a user's information.

Transfers and actions

Now, let's talk about two crucial concepts for interacting with and managing assets and states within smart contracts – **transfers** and **actions**:

*   Transfers refer to the movement of tokens (NEAR tokens) from one account to another. Transfers enable value exchange and facilitate transactions within the blockchain. Developers can initiate transfers programmatically within smart contracts using the NEAR SDK. Here's a simple code example:

```
use near_sdk::ext_contract;
use near_sdk::near_bindgen;
use near_sdk::AccountId;
#[near_bindgen]
pub struct MyContract {}
#[ext_contract(near_token_ext)]
pub trait NEARToken {
```

```
        fn transfer(&mut self, receiver_id: String, amount: u128);
    }
    impl MyContract {
        pub fn transfer_tokens(&mut self, receiver_id: String,
    amount: u128) {
            let contract_account_id: AccountId = todo!();
            near_token_ext::ext(contract_account_id).
    transfer(receiver_id, amount);
        }
    }
    fn main() {}
```

In this example, the `transfer_tokens` function initiates a transfer of NEAR tokens to the specified `receiver_id`.

- Actions represent operations that modify the state or trigger specific behaviors within smart contracts. Actions encapsulate a specific intent or functionality within a contract and can be invoked by external entities or other contract methods. Here's a code example:

```
    use near_sdk::env;
    use near_sdk::near_bindgen;

    #[near_bindgen]
    pub struct MyContract {
        counter: u32,
    }

    impl MyContract {
        pub fn increment_counter(&mut self) {
            self.counter += 1;
        }

        pub fn get_counter(&self) -> u32 {
            self.counter
        }
    }

    fn main() {}
```

In this example, the `increment_counter` action increases the value of the counter by one, and `get_counter` retrieves the current value of the counter.

Cross contract calls

Cross-contract calls allow smart contracts to interact and invoke methods on other contracts. This functionality enables contracts to collaborate, exchange data, and trigger actions across the NEAR blockchain. Let's explore cross-contract calls through code examples.

To initiate a cross-contract call, the NEAR SDK provides the `Promise` struct, which allows contracts to invoke methods on other contracts asynchronously. Here's an example:

```
use near_sdk::{env, near_bindgen, AccountId, Promise};
#[near_bindgen]
pub struct ContractA {}
#[near_bindgen]
impl ContractA {
    pub fn call_contract_b(&self, account_id: AccountId, amount: u128)
{
        let promise = Promise::new(account_id).function_call(
            "do_something".to_owned(),
            vec![],
            amount,
            env::prepaid_gas() - near_sdk::Gas(10),
        );
        promise.as_return();
    }
}
```

In this example, `ContractA` initiates a cross-contract call to `ContractB` by invoking the `do_something` method. `Promise::new` creates a new promise with the target account ID, and `function_call` specifies the method and its arguments. The promise is then executed asynchronously using the `then` function.

On the receiving contract's side, `ContractB`, the invoked method would look like this:

```
use near_sdk::{env, near_bindgen};

#[near_bindgen]
pub struct ContractB {}

#[near_bindgen]
impl ContractB {
    pub fn do_something(&self) {
        // Perform some action
        env::log_str(b"Doing something...");
    }
}
```

In this example, `ContractB` receives the cross-contract call and executes the `do_something` method, which logs a message using `env::log`.

NEAR CLI deep dive

Now, let's explore the NEAR CLI, providing insight into how it streamlines the process of developing and interacting with smart contracts on the NEAR blockchain.

The NEAR CLI enables us to build smart contracts easily and also perform operations around them through a simple CLI tool. We're specifically learning about these commands because they form the foundation for creating, deploying, and interacting with smart contracts on NEAR. You can explore the complete set of NEAR CLI commands in the official documentation (`https://docs.near.org/tools/near-cli-rs`). Here are some of the commands:

- `near init`: This command initializes a new NEAR project in the current directory, creating configuration files and setting up the necessary project structure.

- `near login`: This command allows you to authenticate yourself with a NEAR account, providing access to your account and enabling transactions and interactions with the NEAR blockchain.

- `near deploy`: With this command, you can deploy a smart contract to the NEAR blockchain. It requires the path to the compiled contract (WASM file) and the account ID of the deploying account.

- `near call`: This command invokes a method on a deployed contract. It requires the contract's account ID, the method name, and any required arguments. It allows you to interact with and retrieve data from a smart contract.

- `near view`: This command is similar to `near call` but is used for read-only operations. It allows you to retrieve data from the contract without modifying the state.

- `near state`: This command fetches and displays the current state of a smart contract. It provides useful information about the contract's storage usage and other details.

We will now start building a project, applying the knowledge and tools we've acquired while exploring the NEAR Protocol's features and development resources. For all commands, access `https://docs.near.org/tools/near-cli`.

Creating our first project with NEAR

In this section, we will explore smart contract development using the NEAR Protocol. Our focus will be on creating a crossword game smart contract while leveraging the capabilities of NEAR to build an interactive and decentralized gaming experience.

Crossword games have always captivated people with their intellectual challenges and the joy of uncovering hidden words. Traditionally played on paper or in digital formats, crossword games have now found their way into the realm of blockchain technology. By building a crossword game smart contract, we can combine the time-tested enjoyment of crossword puzzles with the security, transparency, and decentralization provided by NEAR.

Let's take a step-by-step journey to develop the crossword game smart contract. We will start by understanding the structure and rules of the crossword game itself.

Understanding the structure and rules of the crossword game

The smart contract we'll build will allow players to create new crossword games, providing a title and a grid of squares. Players can fill in the blanks with the correct letters to form words both horizontally and vertically. The contract will validate the submissions, maintain the game state, and allow players to interact with the game by submitting words.

To facilitate this, we will define methods within the smart contract, such as creating a new game, retrieving game details, and submitting words. These methods will handle the game's logic and ensure fair play. We will also define a struct, CrosswordGame, to represent the state of an individual game, including the title, grid, and other relevant fields. We will then dive into the code's implementation, using Rust and the NEAR SDK to construct the smart contract.

Next, we will set up our development environment, ensuring that we have the necessary tools and accounts to proceed.

Setting up the development environment

To begin, let's ensure that you have the necessary tools and setup in place for smart contract development on NEAR. Here's a quick checklist:

1. **Install the NEAR CLI**: The NEAR CLI provides a CLI for interacting with the NEAR Protocol. We covered the installation for macOS, Ubuntu, and Windows in the *Installation* section.

2. **Create a NEAR account**: Use the NEAR CLI to create a NEAR account that will serve as the contract deployer and game administrator. Open your terminal or command prompt and run the following command to create a new NEAR account:

   ```
   near create-account <your_account_id> --useFaucet
   ```

3. Replace your_account_id with the desired name for your NEAR account:

   I.   Follow the instructions provided by the NEAR CLI. You'll need to choose a unique account ID and create a password for your account.

   II.   Once the account creation process is complete, make a note of your account ID. You'll use it for deploying and interacting with your smart contract.

4. **Set up a local testnet**: For development and testing purposes, set up a local *testnet* using the NEAR CLI. This will allow you to deploy and test your smart contract locally before deploying it to the main network. To set up a local testnet for development and testing purposes, follow these steps:

    I.   Open your terminal or command prompt and run the `nearup` command to initialize the NEAR testnet.

    II.   Once the testnet is up and running, you can deploy and test your smart contract locally before deploying it to the main network.

    III. To deploy your smart contract to the local testnet, navigate to your project directory in the terminal and run the following command:

    ```
    near deploy --accountId your_account_id --wasmFile path_to_
    wasm_file
    ```

    IV.   Replace `your_account_id` with your NEAR account ID and `path_to_wasm_file` with the path to your compiled smart contract's `.wasm` file.

    V.   You can now interact with your smart contract on the local testnet using the NEAR CLI or SDK libraries.

Now that our development environment is ready, let's dive into the actual smart contract implementation.

Creating a smart contract skeleton

The smart contract we'll build will allow players to create new crossword games, providing a title and a grid of squares. Players can fill in the blanks with the correct letters to form words both horizontally and vertically. The contract will validate the submissions, maintain the game state, and allow players to interact with the game by submitting words.

To facilitate this, we will define methods within the smart contract, such as creating a new game, retrieving game details, and submitting words. These methods will handle the game's logic and ensure fair play. We will also define a struct, `CrosswordGame`, to represent the state of an individual game, including the title, grid, and other relevant fields.

We will use Rust and the NEAR SDK to build our crossword game smart contract while following these steps:

1. Using the NEAR CLI, create a new project named `crossword-game-contract` with the following command:

    ```
    npx create-near-app@latest
    ```

2.   After running this command, you need to select **A Near Smart Contract**, as shown in *Figure 8.5*:

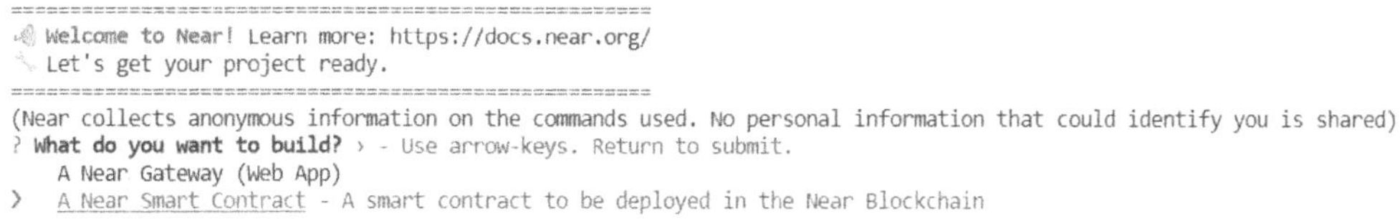

```
=====================================================
Welcome to Near! Learn more: https://docs.near.org/
Let's get your project ready.
=====================================================
(Near collects anonymous information on the commands used. No personal information that could identify you is shared)
? What do you want to build? › - Use arrow-keys. Return to submit.
    A Near Gateway (Web App)
›   A Near Smart Contract - A smart contract to be deployed in the Near Blockchain
```

Figure 8.5 – A Near Smart Contract

3.   Now, select the **Rust Contract** option (see *Figure 8.6*):

```
=====================================================
Welcome to Near! Learn more: https://docs.near.org/
Let's get your project ready.
=====================================================
(Near collects anonymous information on the commands used. No personal information that could identify you is shared)
✓ What do you want to build? › A Near Smart Contract
? Select a smart contract template for your project › - Use arrow-keys. Return to submit.
    JS/TS Contract
›   Rust Contract - A Near contract written in Rust
```

Figure 8.6 – Rust Contract

4.   After that, select **Tests written in Rust** (see *Figure 8.7*):

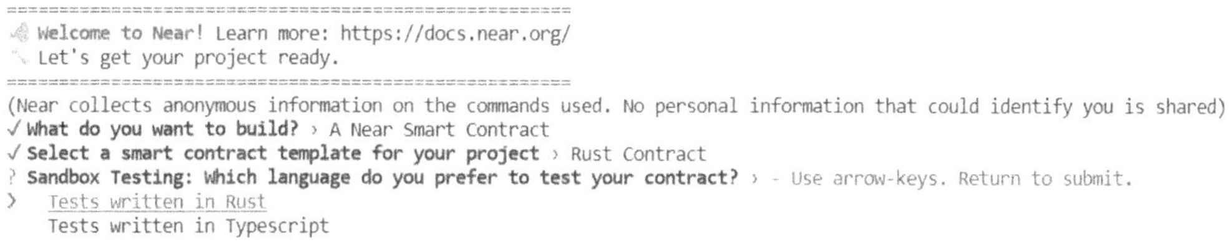

```
=====================================================
Welcome to Near! Learn more: https://docs.near.org/
Let's get your project ready.
=====================================================
(Near collects anonymous information on the commands used. No personal information that could identify you is shared)
✓ What do you want to build? › A Near Smart Contract
✓ Select a smart contract template for your project › Rust Contract
? Sandbox Testing: Which language do you prefer to test your contract? › - Use arrow-keys. Return to submit.
›   Tests written in Rust
    Tests written in Typescript
```

Figure 8.7 – Rust tests

5.   Now, you can name the contract `crossword-game-contract` (see *Figure 8.8*):

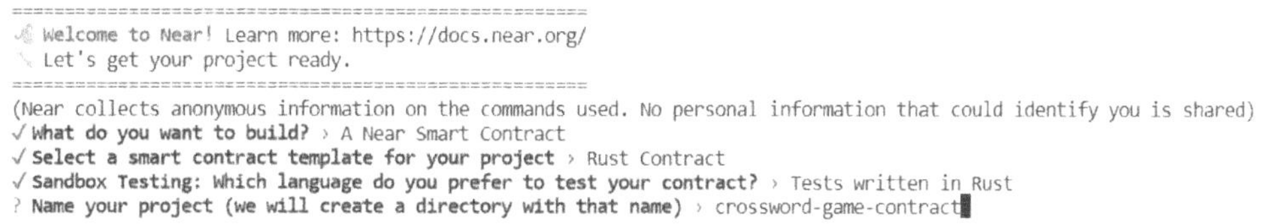

```
=====================================================
Welcome to Near! Learn more: https://docs.near.org/
Let's get your project ready.
=====================================================
(Near collects anonymous information on the commands used. No personal information that could identify you is shared)
✓ What do you want to build? › A Near Smart Contract
✓ Select a smart contract template for your project › Rust Contract
✓ Sandbox Testing: Which language do you prefer to test your contract? › Tests written in Rust
? Name your project (we will create a directory with that name) › crossword-game-contract
```

Figure 8.8 – Rust Contract

6.   Replace `your_account_id` with the NEAR account you created earlier.

7. Explore the structure:

    - Inside the crossword-game-contract directory, you will find the structure of a basic NEAR smart contract project. Here are some of the key files and directories:

      - src/lib.rs: This is the main entry point for our smart contract. It contains the contract's logic and associated methods.

      - Cargo.toml: This file specifies the project's dependencies and settings.

8. Implement the crossword game logic:

    - Open src/lib.rs in your preferred code editor. This is where we will define the contract's logic using Rust and the NEAR SDK:

```
// Import necessary NEAR SDK modules
[dependencies]
serde = { version = "1.0", features = ["derive"] }

use near_sdk::borsh::{self, BorshDeserialize, BorshSerialize};
use near_sdk::{env, near_bindgen, ext_contract, Promise};
// Define the contract struct
#[near_bindgen]
#[derive(Default, BorshDeserialize, BorshSerialize)]
pub struct CrosswordGameContract {
    // Add contract state variables here
}
#[near_bindgen]
use serde::{Serialize, Deserialize};
impl CrosswordGameContract {
    // Add contract methods here
}
```

    - In the preceding code snippet, we import the required modules from the NEAR SDK and define the CrosswordGameContract struct, which will hold the contract's state variables. We also define the contract's entry point using the near_bindgen attribute.

9. Now, let's add some methods to our smart contract to handle game-related operations:

```
impl CrosswordGameContract {
    // Create a new crossword game
    pub fn create_game(&mut self, title: String, grid:
Vec<Vec<Option<u8>>>) {
        // Implement game creation logic here
    }
    // Get the crossword game details
```

```
        pub fn get_game(&self, game_id: u64) ->
    Option<CrosswordGame> {
            // Implement game retrieval logic here
        }
        // Submit a word for a specific game
        pub fn submit_word(&mut self, game_id: u64, word: String) ->
    bool {
            // Implement word submission logic here
        }
    }
```

In the preceding code snippet, we define three methods: `create_game`, `get_game`, and `submit_word`. These methods handle the creation of new games, retrieval of game details, and submission of words, respectively.

10. Finally, we must define the `CrosswordGame` struct so that it represents the state of an individual crossword game:

```
#[derive(BorshDeserialize, BorshSerialize)]
pub struct CrosswordGame {
    pub title: String,
    pub grid: Vec<Vec<Option<u8>>>,
    // Add any other required game-related fields
}
```

In this snippet, we define `CrosswordGame` as a struct that holds the game's title, grid, and any other necessary fields.

Now that the crossword game smart contract is complete, we will proceed to test it locally, ensuring that it functions as intended.

Testing and deployment

To test and deploy the crossword game smart contract, we can follow these steps:

1. To test the contract locally, run the following command in your project's root directory:

```
cargo ./test.sh -- --nocapture
```

This command executes the tests defined in the `tests` directory of your project and displays the output.

2. To deploy the contract to the NEAR testnet, run the following command:

```
near ./deploy.sh --account_id <your_account_id> --wasm_file
target/wasm32-unknown-unknown/release/crossword_game_contract.
wasm
```

3. Replace `your_account_id` with your NEAR account ID. This command will compile your contract and deploy it to the specified account.

Now that we have deployed our crossword game smart contract successfully, let's explore how to interact with it.

Interacting with the contract

Congratulations! You have successfully built and deployed your crossword game smart contract with the NEAR Protocol. Now, let's explore how to interact with it.

To interact with the deployed smart contract from a client application, you can use the NEAR SDK libraries that are available for different programming languages. These libraries provide convenient functions to call contract methods, fetch contract state, and handle transactions.

Here's an example of how you can use the NEAR JavaScript SDK to interact with the deployed contract:

```
const near = require('near-api-js');

async function interactWithCrosswordGameContract() {
    const keyStore = new near.keyStores.InMemoryKeyStore();
    const nearConfig = {
        keyStore,
        nodeUrl: 'https://rpc.testnet.near.org',
        networkId: 'testnet',
        contractName: 'your_account_id',
    };
    const near = await near.connect(nearConfig);

    const account = await near.account(nearConfig.contractName);
    const crosswordGameContract = new near.Contract(
        account,
        nearConfig.contractName,
        {
            viewMethods: ['get_game'],
            changeMethods: ['create_game', 'submit_word'],
            sender: nearConfig.contractName,
        }
    );

    // Interact with the contract methods
    const gameDetails = await crosswordGameContract.get_game({ game_
id: 1 });
    console.log('Game Details:', gameDetails);

    // Add more contract interactions here
}

interactWithCrosswordGameContract();
```

In this code snippet, we initialize the NEAR JavaScript SDK, configure the NEAR network, and create an instance of `CrosswordGameContract` using the contract's `view` and `change` methods. We can then call these methods to interact with the contract.

Summary

In this chapter, we delved into the NEAR Protocol, a blockchain platform tailored for developers, and explored its foundational concepts. We discussed essential elements, such as accounts, access keys, smart contracts, and blockchain state, establishing a solid understanding of NEAR's core principles. We highlighted NEAR's developer-friendly ecosystem, which features implicit accounts, named accounts, full access keys, and function keys, making it an attractive choice for blockchain projects.

We also ventured into advanced topics, including transactions, gas, data flow, tokens, storage options, validators, and consensus mechanisms. This deep dive into NEAR's intricacies demonstrated its robust capabilities for building dApps. Finally, we got hands-on experience by installing the NEAR SDK and NEAR CLI, creating a crossword game smart contract, and deploying it, illustrating the practicality and versatility of NEAR in application development.

In the upcoming chapter, we will delve deeper into the Polkadot ecosystem, examining the fundamental components of Polkadot, Kusama, and Substrate. We will start by unraveling the intricate architecture of Polkadot, shedding light on its relay chain, parachains, and shared security model, which underpin its remarkable capabilities in enabling cross-chain data and asset transfers.

Part 4:
Polkadot and Substrate

In this part, we will build our own custom blockchain, but this time, not from scratch but with a dedicated framework for building blockchains called Substrate, which makes it super simple to build our own chains, customize them, and also deploy them.

This part has the following chapters:

9
Exploring Polkadot, Kusama, and Substrate

In this chapter, we will dedicate our efforts to exploring **Polkadot**, **Kusama**, and **Substrate** – the three interconnected components that form the foundation of the **Polkadot** ecosystem. In the context of the rapidly evolving world of blockchain technology, understanding these components is of paramount importance. Our journey in this chapter will take us through a comprehensive exploration of various facets, ensuring that you grasp not only the fundamental concepts but also the practical applications and interactions within the Polkadot ecosystem.

Throughout this chapter, we will delve into a wide range of topics that are essential for anyone seeking a solid grasp of Polkadot, Kusama, and Substrate. By the end of our journey, you can expect to gain a comprehensive understanding of the Polkadot ecosystem. We will be covering the following topics in this chapter:

- Introducing Polkadot
- Understanding the core concepts of Polkadot
- Learning about Kusama
- Introducing Substrate

Introducing Polkadot

Polkadot is an innovative blockchain network that has gained significant attention in the cryptocurrency and **decentralized application (dApp)** development space. Traditional blockchains face scalability, interoperability, security, upgradeability, and innovation challenges. Polkadot addresses these by using a multi-chain structure with parallel parachains, enabling higher scalability and interoperability. Shared security minimizes vulnerabilities, and its upgrade mechanism allows for seamless updates. The modular design of Polkadot encourages innovation by facilitating rapid development and deployment of new features on custom parachains.

At its core, Polkadot utilizes a unique multi-chain architecture (*Figure 9.1*) that consists of a central **relay chain** and multiple parallel chains called **parachains**:

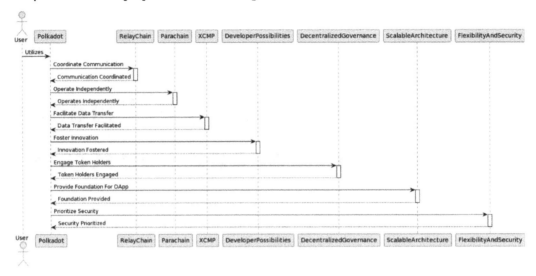

Figure 9. 1 – Illustrating the interaction between the main components and features of Polkadot

The relay chain acts as the backbone of the network, coordinating the communication and consensus among the parachains. Parachains operate independently, allowing for custom configurations and specialized functionalities while benefiting from the shared security and interoperability provided by the Polkadot network.

Let's discuss some of the key features of PolkaDot:

- **Interoperability through XCM**: Polkadot facilitates seamless data transfer and asset exchange between parachains using the cross-chain message passing protocol known as XCM, promoting interoperability among diverse blockchains

- **Developer possibilities**: Interoperability empowers developers to combine distinct blockchain features, fostering innovation and utilizing varied capabilities within a unified network

- **Decentralized governance**: Polkadot's governance model engages token holders in decision-making and network upgrades, ensuring a decentralized and community-driven evolution

- **Scalable architecture**: With a scalable structure, Polkadot offers a strong foundation for dApp development across industries

- **Flexibility and security**: Polkadot's design emphasizes security and flexibility, positioning it as a significant blockchain ecosystem contributor that advances decentralized technologies

Interoperability is central to Polkadot's vision, and XCM is used for achieving interoperability. We will delve into the details of XCM later in this chapter.

First, let's learn about interoperability.

Interoperability

Interoperability lies at the core of the Polkadot network, revolutionizing the blockchain landscape by enabling seamless communication and collaboration between disparate blockchains.

At the heart of Polkadot's interoperability is the relay chain, which serves as the hub connecting different parachains and facilitating cross-chain communication. Parachains can communicate with each other and with external blockchains through the relay chain thanks to the XCM protocol. This allows for the secure transfer of assets, data, and information across various chains within the Polkadot network and beyond.

Polkadot's interoperability extends beyond its network through the use of **bridges**. These bridges establish connections between Polkadot and external blockchains, such as Ethereum or Bitcoin, enabling the transfer of assets and data between different ecosystems. Bridges leverage interoperability protocols and technologies to ensure seamless interoperability and expand the possibilities for cross-chain collaboration.

By facilitating interoperability, Polkadot opens up new avenues for dApp development.

Figure 9.2 shows how relay chains, parachains, parathreads, and bridges work together smoothly. It illustrates the strong network setup of the system:

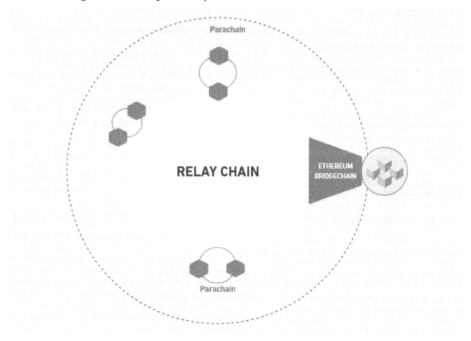

Figure 9. 2 – Interoperability between relay chains, parachains, parathreads, and bridges

Figure 9.2 illustrates Polkadot's unique interoperability. Relay chains act as the backbone, connecting various specialized chains called parachains and more flexible parathreads. Bridges extend connectivity beyond Polkadot, enabling seamless communication with other blockchains. This interconnected structure fosters a powerful and scalable ecosystem for different types of blockchain applications.

Relay chain

The **relay chain** is a fundamental component of the Polkadot network, serving as the central hub that connects and secures the entire ecosystem. It plays a vital role in coordinating and facilitating communication between the different **parachains** in the network.

One of the key features of the relay chain is its ability to facilitate interoperability among parachains. It enables seamless data transfer, asset exchange, and even the execution of transactions between different parachains. This interoperability is achieved through the innovative XCM protocol, which we will learn more about in the *XCM section*. This allows for the efficient and secure exchange of information across the network.

Furthermore, the relay chain serves as the backbone of the Polkadot network's governance framework. This framework incorporates the use of **DOT tokens**, the native cryptocurrency of Polkadot. DOT token holders actively engage in decision-making processes by staking their tokens and participating in voting on **referenda proposals**. These referenda proposals encompass key network changes, upgrades, and policies. This decentralized governance mechanism underscores the network's commitment to transparency and community-driven evolution.

The relay chain's **Nominated Proof of Stake** (**NPoS**) consensus mechanism and shared security model collaboratively bolster Polkadot's security and resilience. Through NPoS, token holders nominate validators to validate transactions and propose blocks, upholding network integrity. The shared security model extends this by allowing specialized blockchains, known as parachains, to leverage the relay chain's security. While the relay chain ensures network-wide security, parachains enable tailored functionalities and optimizations for specific use cases, enhancing the network's versatility. This dual approach ensures validated transactions, finalized blocks, and ecosystem integrity, underscoring Polkadot's commitment to security and adaptability.

Parachains play a crucial role in realizing the vision of a scalable and interoperable blockchain ecosystem. Parachains are independent chains that run in parallel to the Polkadot relay chain, each with its own set of validators and customizable features.

Parachains enable developers to create specialized blockchains that cater to specific use cases and requirements. They can be designed with different consensus mechanisms, governance models, and functionalities, allowing for a high degree of customization. This flexibility makes Polkadot a versatile platform for building a wide range of dApps.

Interoperability is a key aspect of the Polkadot network, and parachains are at the forefront of enabling seamless communication and data transfer between different chains. Parachains can communicate with each other and with the relay chain using the XCM protocol. This allows information, assets, and even the execution of transactions to be exchanged across parachains, fostering a highly interconnected ecosystem.

Parachains benefit from the shared security model of Polkadot. The security of each parachain is enhanced by the collective security of the entire network (we will learn more about this in the *Shared security* section), which is provided by the validators of the relay chain. This shared security model ensures the integrity and reliability of the parachains, bolstering trust and resilience in the network.

Moving forward, we will delve into another vital aspect of Polkadot's architecture: **parathreads**. This innovative concept offers an alternative approach to *parachains*, providing more flexible and cost-effective options for securing a slot on the relay chain.

Parathreads

Parathreads offer a flexible and cost-effective solution for blockchain projects that do not require continuous and full-time access to a parachain slot. Parathreads allow these projects to gain intermittent access to the Polkadot network, balancing their needs for scalability and interoperability with a more economical approach.

Unlike parachains, which have dedicated slots on the relay chain, parathreads do not occupy a fixed slot permanently. Instead, they can dynamically acquire slots on-demand when needed. This enables projects to pay for the exact usage they require, making it a cost-efficient option for blockchain applications with sporadic or varying demand patterns.

Parathreads provide the benefits of interoperability and security within the Polkadot network. They can communicate and interact with both parachains and other parathreads through the XCM protocol, allowing for seamless data transfer and collaboration. They also benefit from the shared security model of Polkadot, leveraging the overall security provided by the validators of the relay chain.

The introduction of parathreads expands the capabilities of the Polkadot network, accommodating a wider range of projects with different resource requirements. It promotes inclusivity and allows for the efficient utilization of network resources, ensuring scalability and flexibility for blockchain applications.

Bridges

Bridges play a crucial role in the Polkadot network by enabling connectivity and interoperability between Polkadot and external blockchains. They serve as vital links that facilitate the transfer of assets, data, and information across different blockchain ecosystems.

Polkadot's architecture allows for the creation of bridges that connect the Polkadot network to other blockchains, such as Ethereum, Bitcoin, or any other compatible blockchain. These bridges establish a secure and reliable communication channel, enabling the seamless movement of assets and information between Polkadot and external networks.

Bridges in the Polkadot network leverage various interoperability protocols and technologies to facilitate cross-chain communication. They enable the transfer of tokens and assets between different blockchain networks, promoting liquidity and composability across disparate ecosystems.

Now, let's explore another integral aspect of the Polkadot ecosystem: accounts. These serve as the foundational units through which users interact and participate in the ecosystem. Polkadot accounts enable users to securely manage their assets, execute transactions, and engage in governance processes within the network.

Accounts

Accounts serve as the foundational units through which users interact and participate in the ecosystem. Polkadot accounts enable users to securely manage their assets, execute transactions, and engage in governance processes within the network.

In the Polkadot network, accounts are identified by unique addresses, similar to other blockchain platforms. These addresses serve as public identifiers associated with a specific account and are used for sending and receiving transactions. Polkadot addresses are derived from the user's cryptographic key pair and provide a secure and verifiable way to authenticate transactions and interactions.

Polkadot's account system is designed to accommodate various user needs and scenarios. Within this system, accounts can be categorized into different types, each serving specific functions:

- **Regular accounts**: Typically owned by individual users and used for managing and transferring assets.

- **Contract accounts**: Accounts associated with smart contracts, allowing for the execution of predefined functions and interactions with the blockchain.

- **Governance accounts**: Special accounts that hold voting power and are used to participate in the governance processes of the network.

- **Multi-signature (multi-sig) accounts**: Polkadot's account system supports multi-sig accounts, which require multiple signatures from different parties to authorize a transaction. This enhances security and mitigates the risk of single-point failures.

Now that we have a clear understanding of the different types of accounts, let's delve into the essential concept of transactions within the Polkadot ecosystem.

Transactions

The Polkadot ecosystem is renowned for its advanced **transaction** capabilities, providing a secure and scalable network for seamless asset transfers. Transactions within the Polkadot network occur through its native cryptocurrency, *DOT*, which serves as a medium of exchange and a governance token.

As shown in the following figure, when a transaction is initiated on the Polkadot network, it undergoes a series of steps. Initially, the transaction is validated by the network's validators, who ascertain its integrity and adherence to the rules defined by the underlying protocol of the Polkadot blockchain. These protocol rules constitute a set of predefined guidelines that dictate the proper functioning of the blockchain network. Once validated, the transaction becomes part of a block that is subsequently added to the relay chain. This relay chain acts as a central hub, orchestrating and finalizing transactions, thereby maintaining the overall coherence and integrity of the Polkadot ecosystem:

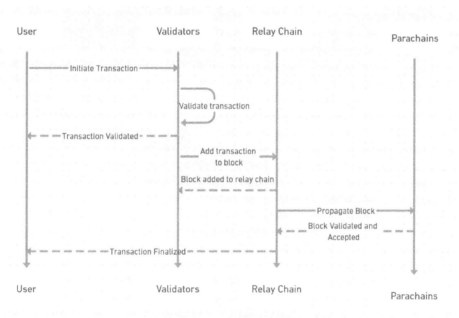

Figure 9. 3 – A transaction getting validated and added to the blockchain

To incentivize validators and ensure network security, Polkadot employs a mechanism called **NPoS**, which we will learn about more in the *NPoS election algorithms* section. Validators are chosen through a staking process, where DOT token holders can nominate trustworthy validators. Validators are then selected based on their reputation and the amount of DOT tokens they have staked. Validators who successfully validate transactions are rewarded with DOT tokens, while those who act maliciously or fail to fulfill their duties may face penalties.

The Polkadot ecosystem also provides users with the flexibility to customize their transaction types based on their specific needs. Developers can create custom transaction types within their parachains, allowing for the execution of complex transactions beyond simple asset transfers.

Creating custom transaction types in a Substrate-based parachain involves defining custom runtime modules and specifying the logic for these transactions. Let's take a look at an example to understand this. The following code provides a simplified example of using Rust and Substrate's **Flexible Runtime**

Aggregation Modular Extension (**FRAME**) framework to define a custom transaction type for a parachain. Let's take a closer look:

1. First, we define a custom transaction struct called `MyCustomTransaction` with fields such as `sender` and `amount`. You can add any other fields that are specific to your use case:

```
// Import necessary dependencies from Substrate

use frame_system::{Module as System, RawOrigin, ensure_signed,
ensure_root};
use frame_system::pallet::Config::{AccountId, Balance}
// Define your custom transaction module
pub mod my_custom_module {
    use super::*;


    // Define the custom transaction struct
    #[derive(codec::Encode, codec::Decode, Default, Clone,
PartialEq)]
    pub struct MyCustomTransaction {
        // Define transaction fields here
        pub sender: AccountId,
        pub amount: Balance,
        // Add any other fields you need

    }
```

2. Then, we implement a dispatchable function called `my_custom_transaction` within the custom module. This function defines the logic for processing your custom transaction. In this example, we ensure the sender's authenticity using `ensure_signed`; you can add your specific logic to process the transaction:

```
use frame_support::{dispatch::DispatchResult, pallet_
prelude::*};
use frame_system::pallet_prelude::*;

pub struct MyCustomTransaction {
    // Add fields for your custom transaction
    // e.g., sender: AccountId, amount: Balance
}


pub trait Config: frame_system::Config {
    // Add any additional configuration types here
    frame_system::Config>::Event>;
}

pub mod pallet {
```

```
        use super::*;

        #[pallet::pallet]
        #[pallet::generate_store(pub(super) trait Store)]

        pub struct Pallet<T>(_);

        #[pallet::call]

        impl<T: Config> Pallet<T> {
    // Implement your custom transaction as a dispatchable function
            #[pallet::weight(0)] // Set the correct weight
            pub fn my_custom_transaction(
                origin: OriginFor<T>,
                transaction: MyCustomTransaction,
            ) -> DispatchResult {
                let sender = ensure_signed(origin)?;
     // Your custom logic for processing the transaction here
     // You can access the fields of transaction, like transaction.
    sender and transaction.amount

                Ok(())
            }
        }
    }
```

3. You can emit events, perform actions, or execute any business logic within the `my_custom_transaction` function, as shown in the preceding code snippet.

4. Finally, ensure that your custom module is included in the runtime configuration of your parachain, as shown in the following code snippet:

```
// Ensure your custom module is included in the runtime
configuration
impl<T: Trait> frame_system::Module<T> {
    fn dispatch_bypass_filter(
        origin: T::Origin,
        _call: T::Call,
    ) -> dispatch::DispatchResult {
        ensure_root(origin)?;
        Ok(())
    }
}
```

Now, let's understand what tokens and assets are.

Tokens and assets

Polkadot operates on a multichain architecture, with each chain having native tokens and assets.

The primary token within the Polkadot ecosystem is **DOT**. DOT serves as the native cryptocurrency of the network and acts as a utility token for governance, staking, and bonding purposes. DOT holders have the power to participate in the network's governance, including voting on proposals and decisions that shape the future of the ecosystem. They decide on upgrades, allocate resources, maintain security, promote innovation, and foster community engagement, ensuring the network's evolution and adherence to decentralized principles.

In addition to DOT, each parachain within the Polkadot network can have its own native token or asset. Parachains are independent blockchains that connect to the Polkadot relay chain and can have their own unique features, governance models, and token economies. These native tokens or assets are used to power the specific functionalities and services offered by each parachain.

The interoperability of the Polkadot network enables the seamless transfer of assets between different parachains. This means that tokens and assets from one parachain can be exchanged or used within another parachain, providing liquidity and expanding the utility of these assets across the ecosystem.

Polkadot also supports the creation of **cross-chain bridges**, which enable the transfer of assets between Polkadot and other blockchain networks. These bridges allow for interoperability with external networks such as Ethereum, enabling the movement of assets between different ecosystems and unlocking new opportunities for collaboration and innovation.

Furthermore, Polkadot offers a framework for the creation of dApps and services. Developers leverage the Polkadot ecosystem to craft their tokens through smart contracts, parachains, or custom-built blockchain modules. These tokens are programmable, allowing developers to define their specific functionalities, such as transaction rules, supply limits, and ownership structures. Smart contracts, when executed on Polkadot's parachains, enable the automation of in-app transactions and user incentives. Additionally, token standards such as ERC-20 or Polkadot's native standards simplify the creation and interoperability of these assets. Tokens can also be linked to off-chain or real-world assets through oracles, further extending their utility in representing ownership rights over physical assets. In essence, Polkadot's versatile infrastructure empowers developers to engineer tokens that serve diverse purposes within their dApps, enhancing the ecosystem's flexibility and innovation potential

Now, it's time to take a deep dive into **non-fungible tokens** (**NFTs**).

NFTs

In the Polkadot ecosystem, NFTs can be created on specific parachains or can be bridged from other networks into Polkadot using cross-chain bridges. NFTs are unique digital assets that represent ownership of one-of-a-kind items, such as digital art, collectibles, or even real-world assets. They provide proof of authenticity and ownership on the blockchain, making them valuable for creators and collectors alike. Just like other components, NFTs can be effortlessly moved and traded between diverse blockchains, thereby boosting liquidity and broadening the accessibility of various digital assets.

In the Polkadot ecosystem, NFTs provide remarkable flexibility as developers can customize their attributes, creating unique qualities such as limited editions or provable scarcity. These adaptable NFTs can represent an extensive range of digital assets, from artwork and collectibles to virtual real estate and in-game items. What's more, Polkadot's infrastructure supports the creation of dApps that leverage NFTs. Developers can build NFT marketplaces, platforms for tokenized assets, and gaming ecosystems where NFTs serve as in-game items or digital representations of real-world assets. Polkadot's scalability, interoperability, and security bolster these applications, offering a robust foundation for NFT-based experiences across diverse use cases.

Additionally, the decentralized nature of Polkadot ensures that NFT ownership and transactions are transparent, secure, and verifiable. The use of blockchain technology guarantees the authenticity and provenance of NFTs, preventing counterfeiting and providing a trustworthy environment for creators and collectors.

Understanding the core concepts of PolkaDot

In this section, we'll dive deep into the Polkadot ecosystem. Here, we will explore a range of topics that are fundamental to understanding the inner workings and innovative features of Polkadot.

We will begin by exploring **XCM**, a revolutionary technology that enables communication and the transfer of assets across different parachains within the Polkadot network. We'll delve into how XCM facilitates interoperability, allowing for seamless transactions and data transfers between interconnected blockchains.

XCM

XCM is a groundbreaking technology within the Polkadot ecosystem that enables seamless communication and asset transfer across different parachains. It plays a pivotal role in Polkadot's interoperability framework by enabling the seamless exchange of messages and data among interconnected blockchains.

Let's gain a deeper understanding of XCM's role in facilitating interoperability within the Polkadot network:

- **Asset and data transfer**: XCM's primary role is to securely and efficiently transfer assets, tokens, and messages between parachains within the Polkadot network, simplifying interactions.

- **Standardized message format**: XCM achieves interoperability through a standardized message format, allowing different parachains to exchange instructions or data, such as token transfers or smart contract execution.

- **Atomic transactions**: XCM ensures atomic transactions, guaranteeing that asset transfers and actions are executed completely or not at all, maintaining the integrity and consistency of transactions across the network.

- **Flexibility**: XCM offers flexibility, enabling parachains to define their own message formats and logic. This customization allows developers to build specialized parachains while retaining the ability to communicate and transfer assets seamlessly with others in the network.

XCM plays a crucial role in unlocking the full potential of the Polkadot ecosystem. It promotes collaboration and interoperability between different blockchains, enabling a vibrant and connected network of dApps and services. With XCM, Polkadot offers a scalable and flexible framework for building the next generation of blockchain solutions that can seamlessly communicate and transact with one another.

Now, let's learn more about how shared security works.

Shared security

The concept of **shared security** within the Polkadot ecosystem is designed to enhance the overall security and resilience of the network. Shared security ensures that multiple parachains within the Polkadot network can collectively benefit from a shared pool of security resources.

Each parachain has its own set of validators responsible for validating transactions and maintaining the security of the respective parachain. However, through shared security, parachains can leverage the collective security provided by the Polkadot relay chain. This is depicted in the following figure:

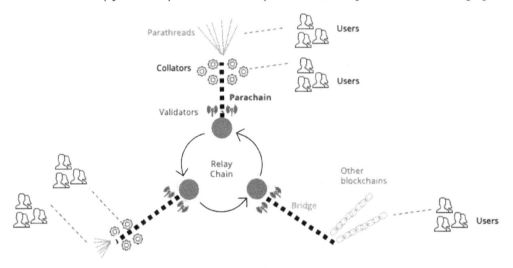

Figure 9.4 – A closer look at the shared security model

The Polkadot relay chain acts as a central hub for coordinating and finalizing transactions across the network. It serves as a security anchor and provides a shared pool of validators that can be allocated to different parachains. By utilizing shared security, parachains can tap into the validator pool of the relay chain, benefiting from its combined computational power and security measures.

Shared security enhances the resilience of the Polkadot ecosystem by distributing the risk and reducing the attack surface. If one parachain faces a security threat or an attack, the shared security model ensures that the other parachains remain protected.

Also, shared security promotes a cooperative environment within the Polkadot network. Validators have the incentive to participate in the security of multiple parachains as it allows them to diversify their stake and earn rewards from multiple sources. This diversification of stake and the opportunity to earn rewards from multiple sources incentivize validators to act in the network's best interest. Validators are motivated to maintain the network's security and stability because their rewards depend on it. By actively participating in consensus and ensuring the integrity of transactions, validators contribute to a collaborative and secure ecosystem as their actions directly impact their potential rewards. This alignment of incentives encourages validators to work collectively toward the network's well-being, enhancing its overall resilience and reliability.

Pallets

Pallets are building blocks that can be added to parachains to extend their capabilities and enable developers to create custom features and applications.

Pallets are like building blocks for parachains in the Polkadot ecosystem, each offering specific functions and features. These functionalities include token standards, smart contract frameworks, governance tools, oracles, and more. Developers can mix and match pallets to create customized parachains tailored to their needs. This modular approach promotes code reuse and streamlines development, making it faster and more efficient to deploy parachains on Polkadot.

For example, consider the following functionalities offered by pallets:

- **Token standards**: Pallets for token standards provide a framework for creating and managing digital assets, enabling developers to issue and manage tokens within their parachains

- **Smart contract frameworks**: These pallets offer the tools and infrastructure for deploying and executing smart contracts, allowing developers to build dApps on Polkadot

- **Governance mechanisms**: Pallets focused on governance enable the implementation of on-chain decision-making processes, helping communities and stakeholders reach consensus and make important network decisions

- **Oracles**: Pallets related to oracles facilitate the integration of external data into the blockchain, enhancing the parachain's ability to interact with the real world by providing reliable off-chain information

These pallets enable developers to assemble parachains with the precise features required for their use cases. Furthermore, pallets enhance Polkadot's interoperability by allowing their use across multiple parachains, enabling seamless communication and asset transfer between chains, and bolstering the ecosystem's versatility.

Next, we'll talk about **staking**.

Staking

Staking is another important process that locks up cryptocurrency tokens as collateral in a blockchain network to support its operations. In return, participants earn rewards or incentives for validating transactions and securing the network.

Staking offers unique features and benefits that contribute to the security, governance, and overall operation of the network. Staking in Polkadot involves locking up DOT tokens to engage in both the network's consensus and governance processes. In the consensus process, validators, who have staked DOT tokens, propose transaction blocks that are then verified and approved by a supermajority of validators. This approval mechanism ensures network security and transaction finality through a two-step process. Participants, including validators and nominators, receive rewards for their involvement, incentivizing them to maintain the network's reliability. However, any misconduct can result in **slashing**, where tokens are forfeited. In essence, Polkadot's staking and consensus system encourages agreement, security, and active network maintenance.

One unique aspect of staking in the Polkadot ecosystem is the ability of token holders to actively participate in network security and governance. By staking DOT tokens, individuals can become validators or nominators and contribute to the network's operations. Tokens are staked by locking them in a designated wallet, committing them as collateral to support the blockchain's security and governance. Validators actively participate in block validation, while nominators select validators to support and share in the rewards generated by the validator's activities. This staking process not only secures the network but also aligns the interests of participants with the network's overall health and performance.

Validators are responsible for validating transactions, producing blocks, and securing the network. They play a crucial role in maintaining the integrity and security of the Polkadot ecosystem. Validators are required to stake a significant amount of DOT tokens as collateral, ensuring their commitment to the network's stability and performance. Becoming a validator typically involves a few key steps:

1. **Staking adequate collateral**: As mentioned previously, validators must stake a significant amount of DOT tokens as collateral. This amount needs to meet or exceed the minimum requirements set by the network.

2. **Technical competence**: Validators need to possess the technical knowledge and infrastructure to maintain and operate a node effectively. This includes hardware reliability, network connectivity, and security measures.

3. **Nominated**: Potential validators may be nominated by token holders who support them by staking their own DOT tokens. Nominators share in the rewards generated by the validators they nominate.

4. **Community trust**: Validators often need to gain the trust of the Polkadot community. This can be achieved by participating in testnets, demonstrating consistent performance, and engaging with the community through transparent communication.

5. **Election**: Validators are elected through Polkadot's NPoS system.

Staking within the Polkadot network involves nominators who delegate their DOT tokens to validators they trust. Nominators follow a multi-step process, starting with the careful selection of validators based on various criteria. These criteria often include a validator's reputation, consistent and reliable performance, the amount of DOT tokens they have committed as collateral, technical competence, and contributions to the network's overall health.

Once selected, nominators delegate their DOT tokens to their chosen validators, typically through the network's user interface or client. By doing so, they actively contribute to the network's security and performance. In return, nominators have the opportunity to earn rewards based on the validators' performance. These rewards are derived from the validators' earnings, which are often a share of the block rewards and transaction fees generated by the network. Overall, nominators play a crucial role in the Polkadot ecosystem by supporting trustworthy validators and reinforcing network reliability while benefiting from the rewards generated by their nominated validators.

A unique feature of staking in the Polkadot ecosystem is the concept of **adaptive staking**. Nominations in the Polkadot network represent a dynamic and participatory aspect of the proof-of-stake system. They allow token holders to choose and support validators they trust and believe will contribute to the network's security and performance. Token holders can adapt their nominations by adjusting the validators they support, adding or removing validators from their list without incurring substantial penalties. This adaptability ensures that token holders have the flexibility to respond to changing conditions, such as shifts in validator performance or reputation. It also promotes a competitive environment among validators, encouraging them to maintain high standards to attract and retain nominations, ultimately enhancing the network's overall security and reliability. It offers flexibility and ensures that token holders can actively participate in network governance by supporting validators aligned with their preferences and values.

Advanced staking concepts

In addition to the basic staking concepts mentioned previously, the Polkadot ecosystem introduces advanced staking concepts that further enhance participation and rewards for stakeholders. These advanced staking concepts provide additional opportunities for engagement and offer unique benefits within the network:

- **Slashing**: Slashing in the Polkadot ecosystem serves as a critical mechanism to deter malicious behavior and uphold network integrity. When a validator or nominator engages in malicious actions or violates network rules, they can face slashing penalties, which involve confiscating or reducing their staked DOT tokens. Examples of malicious actions and rule violations include double-signing, where a validator signs conflicting messages, and equivocation, which involves validators confirming two contradictory blocks. Such actions undermine network security and reliability. The specific algorithms and criteria for determining the extent of confiscation or reduction depend on the nature and severity of the violation, and these rules are typically defined and enforced through Polkadot's governance processes. These mechanisms are designed to discourage malicious behavior, ensuring the overall trustworthiness and security of the network.

- **Validator performance monitoring**: Validator performance in the Polkadot ecosystem is assessed through a set of key criteria that encompass various aspects of their operations. These criteria typically include the following:

  - **Uptime**: This measures how long a validator's node is online and actively participating in block validation. High uptime indicates reliability.

  - **Responsiveness**: Validators should promptly respond to network requests and actions, ensuring the smooth functioning of the network.

  - **Validation accuracy**: Validators are expected to validate transactions and blocks accurately, without producing conflicting or invalid results.

  - **Slashable offenses**: Any malicious or rule-violating behavior, such as double-signing or equivocation, is heavily penalized and impacts a validator's performance.

  - **Community trust**: Validators often earn trust based on their transparency, communication, and contributions to the community.

  - **Nominator support**: The number of DOT tokens staked by nominators supporting a validator can also be an indicator of performance and trustworthiness. These criteria are typically assessed over a specific timeframe, and validators meeting the network's defined standards continue to operate securely. Validators failing to meet these standards may face penalties, and in severe cases, they could be removed from the validator set. This rigorous performance monitoring ensures that the Polkadot network maintains a high level of security and reliability.

- **Reward distribution**: Stakers in the Polkadot ecosystem are rewarded for their participation and contribution to the network. Rewards are distributed based on factors such as the amount of staked tokens, the length of the staking period, and the performance of the validator. This incentivizes stakeholders to actively engage in staking and ensures a fair distribution of rewards.

- **Stash accounts and controller accounts**: In Polkadot, stakeholders have the option to separate their staked DOT tokens into stash accounts and controller accounts. The stash account holds the staked tokens and remains secure, while the controller account manages the staking activities and interacts with the network. This separation provides an added layer of security and control for stakeholders.

- **Validator rotation**: The Polkadot ecosystem employs a validator rotation mechanism to ensure fairness and decentralization. Validators are periodically rotated, allowing different participants to have the opportunity to validate transactions and earn rewards. Validator rotation promotes a dynamic and diverse validator set, contributing to the overall security and resilience of the network.

These advanced staking concepts in the Polkadot ecosystem highlight the platform's commitment to the security, fairness, and active participation of stakeholders. By implementing mechanisms such as slashing, performance monitoring, and reward distribution, Polkadot encourages a robust and engaging staking environment where participants can contribute to network security and governance while being rewarded for their contributions.

Next, we'll explore the main actors within the Polkadot ecosystem.

Main actors

In this section, we'll delve into the roles and responsibilities of three key participants: validators, nominators, and collators (*Figure 9.5*). These actors play crucial roles in ensuring the security, governance, and functionality of the Polkadot network. We will examine their unique contributions, incentives, and how they interact with each other to create a robust and decentralized ecosystem:

Nominators

Secure the Relay Chain by selecting trustworthy validators and staking dots.

Validators

Secure the relay chain by staking dots, validating proofs from collators and participating in consensus with other validators.

Collators

Maintain shards by collecting shard transactions from users and producing proofs for validator.

Figure 9.5 – Consensus roles in Polkadot

We'll start by looking at validators.

Validators

Validators are responsible for validating transactions, producing blocks, and ensuring the security and integrity of the network. They play a crucial role in maintaining consensus and preventing malicious activities within the blockchain.

They are required to stake a significant amount of DOT tokens as collateral, demonstrating their commitment to the network's stability and security. By staking their tokens, validators have a financial stake in the system, aligning their incentives with the successful operation of the network.

Validators are selected through a unique election process in Polkadot known as *NPoS*. This algorithm takes into account various factors, including the amount of stake, reputation, and the number of nominations, to determine the set of active validators. This ensures a diverse and secure validator set that represents the interests of the network participants.

Validators are responsible for verifying transactions, executing smart contracts, and producing blocks that are added to the Polkadot relay chains.

Nominators

Nominators play a vital role in contributing to network security and governance by selecting and supporting validators. As stakeholders, nominators hold DOT tokens and have the ability to delegate their stake to validators of their choice.

By nominating validators, nominators help to secure the network and ensure the integrity of transactions. Nominators assess validators based on reputation, performance, and the amount of stake they have committed. Validators' reputation in Polkadot is typically assessed based on their historical performance, community engagement, and transparency. A strong track record of consistent uptime, accurate validation, and reliability builds a positive reputation. Active involvement in the Polkadot community, transparent communication, and contributions to network development enhance trust. Technical competence and a clean record, free from any slashable offenses or malicious behavior, also contribute to a validator's positive standing. Collectively, these factors establish a validator's reputation as trustworthy and dedicated to maintaining the network's stability and security. They aim to select validators who have a proven track record of reliable validation and contribute to the overall stability of the network.

Nominators participate in the staking process by bonding their tokens to the validators they nominate. This bonding mechanism aligns the interests of nominators with the successful operation of the network as they have a financial stake in the validators they support. In return for their participation, nominators receive a portion of the rewards earned by the validators they nominate.

The role of nominators goes beyond simply staking their tokens. They actively monitor the performance of validators and may adjust their nominations based on changes in performance or reputation. This flexibility allows nominators to make informed decisions and adapt to the evolving dynamics of the network.

Nominators contribute to the decentralization and security of the Polkadot ecosystem by diversifying stakes across multiple validators. This helps to prevent the centralization of power and reduces the risk of any single validator compromising the network.

Collators

Collators are important participants in the Polkadot ecosystem, specifically within the Substrate framework, which we will learn more about in the *Substrate* section. They play a critical role in facilitating the execution and validation of transactions within parachains.

Let's delve into the essential responsibilities of collators in the Polkadot ecosystem, outlining their pivotal role in ensuring seamless transaction processing and enhancing network scalability:

- **Collator responsibilities**: Collators are tasked with collecting, validating, and organizing transactions within their designated parachains.
- **Transaction packaging**: They assemble these transactions into blocks, which are then sent to validators for the final validation and inclusion in the Polkadot relay chain.

- **Intermediaries**: Collators act as intermediaries who bridge the gap between parachains and the Polkadot relay chain, ensuring transactions comply with their specific chain's rules and logic.

- **Scalability enhancement**: Collators play a pivotal role in enhancing the scalability of the Polkadot ecosystem. By handling initial transaction validation and block creation within parachains, they contribute to network efficiency and relieve validators, leading to increased transaction throughput.

- **Proofs of validity**: In addition to their transaction processing duties, collators are also responsible for producing proofs of validity. These proofs provide evidence that the transactions included in the blocks are valid and can be relied upon for final validation.

Collators are incentivized to perform their role diligently through rewards. They receive compensation for their services in the form of transaction fees and block rewards.

NPoS election algorithms

In the Polkadot ecosystem, **NPoS election algorithms** are instrumental in determining which validators and nominators actively participate in network consensus and governance. These algorithms ensure a fair and decentralized process for selecting key actors within the network.

The NPoS election algorithms take various factors into account, including stake, reputation, and the number of nominations received by validators. Validators with higher stakes and more nominations have a greater chance of being selected to participate in block production and transaction validation.

Let's explore the steps involved in NPoS election algorithms:

1. **Era-based elections**: The election process in the Polkadot network operates in eras, which are fixed periods. During each era, a new set of validators is selected based on various factors.

2. **Predefined parameters**: The duration of each era and the number of validators to be elected are predefined and determined through the network's governance mechanisms. These parameters can be adjusted through community governance.

3. **Diverse validator selection**: NPoS algorithms aim to create a diverse and secure validator set. They consider factors beyond stake alone to prevent excessive centralization. This ensures validators represent a broad range of interests in the ecosystem.

4. **Inclusion of reputation**: Reputation is a crucial factor in validator selection. Validators are incentivized to maintain a positive reputation by providing reliable and trustworthy services to the network.

5. **Nominator participation**: Nominators, who delegate their tokens to validators, play a pivotal role. Nominations contribute to a validator's chance of being selected. Nominators can choose validators they trust and support, promoting active participation and reinforcing network security.

6. **Balanced selection**: The NPoS election algorithms strike a balance between stake-based selection and community-driven governance. This balanced approach ensures network security, encourages participation, and upholds the integrity of the Polkadot ecosystem.

Overall, NPoS algorithms create a dynamic and fair process for selecting validators and nominators, enhancing network security and decentralization while actively involving the community in governance decisions.

Next, we will turn our attention to Kusama, Polkadot's experimental network designed to foster innovation and rapid iteration. We will explore Kusama's role as a canary network, where developers can test new features, experiment with governance, and push the boundaries of what's possible in a live environment.

Learning about Kusama

Kusama is a vibrant and experimental environment within the Polkadot ecosystem. In this section, we will delve into the origins of Kusama, its unique benefits, and the key features that make it a fascinating blockchain network.

It is a decentralized blockchain network and a sister network of Polkadot. Created by the same team behind Polkadot, Kusama serves as a canary network designed for experimentation and innovation. It allows developers to test new features, economic models, and governance mechanisms before deploying them on the more stable Polkadot network.

Kusama offers several key benefits that make it an attractive and unique blockchain network within the Polkadot ecosystem:

- **Fast iteration**: Kusama excels in rapid project development and testing. Developers can quickly experiment and refine their ideas with frequent upgrades and a flexible approach, speeding up innovation.

- **Low economic barriers**: Kusama offers a cost-effective entry point. Acquiring Kusama tokens (KSM) and participating in the network is generally more affordable than on Polkadot, fostering a diverse developer and user base.

- **Advanced technology**: Kusama serves as a playground for cutting-edge blockchain technology. Developers benefit from early access to experimental features and solutions, staying at the forefront of innovation.

- **Open governance**: Kusama's transparent governance model invites active involvement from token holders. They can propose and vote on upgrades and changes, ensuring community-driven decision-making and network evolution.

Another crucial aspect we will explore is governance and on-chain upgrades in Kusama. You will understand how the network empowers token holders to actively participate in decision-making processes and propose and vote on upgrades and changes. We will discuss the self-upgradable nature of Kusama and the community-driven governance model that shapes its evolution.

Governance and on-chain upgrades

Kusama's governance model empowers token holders to propose, discuss, and vote on vital network decisions, ensuring inclusivity and active participation:

- **Voting process**: Using Kusama's on-chain governance, proposals undergo voting, where token holders stake KSM tokens to support or oppose them, ensuring decentralized, transparent decision-making

- **Efficient upgrades**: On-chain upgrades bolster flexibility, allowing seamless improvements without hard forks or manual updates, making the process smoother

- **Resilience and democracy**: Kusama's governance fosters openness, inclusivity, and decentralized decision-making, enhancing network resilience, adaptability, and democratic values

Kusama's voting mechanism is pivotal in enabling democratic decision-making. Proposals are introduced and subsequently subjected to a transparent voting process. Token holders, as per their staked KSM tokens, participate by either supporting or opposing these proposals. The weight of their votes is directly proportional to the number of tokens staked, providing a fair representation of the community's preferences. Kusama's voting process ensures that the majority's decision guides the implementation of changes, securing the network's democratic ethos while fostering inclusivity, transparency, and decentralized governance.

Next, we'll delve into the concept of chaos and experimentation within Kusama. We will explore how Kusama's environment encourages risk-taking, innovation, and the testing of new features, economic models, and governance mechanisms before they're deployed on the more stable Polkadot network.

Chaos and experimentation

Chaos and experimentation are fundamental principles of the Kusama network, distinguishing it as a platform that encourages risk-taking and innovation. Kusama's *Expect Chaos* ethos creates an environment where developers and participants can freely experiment with bleeding-edge technologies, economic models, and governance mechanisms.

The chaotic nature of Kusama is intentional and serves a crucial purpose. By design, Kusama allows for rapid iterations and testing of new ideas before they are deployed on the more stable Polkadot network. This approach enables developers to identify and address any potential issues or vulnerabilities, ensuring the highest level of security and stability on Polkadot.

Kusama's chaos and experimentation foster an atmosphere of innovation and freedom. Developers are encouraged to explore unconventional ideas and challenge the status quo. This freedom allows for the exploration of radical solutions and the potential discovery of groundbreaking advancements in decentralized technology.

Participants in the Kusama network understand and embrace the inherent risks involved. They actively engage in testing and providing feedback, contributing to the collective learning and improvement of the ecosystem. The ability to experiment without fear of disrupting the main network creates a dynamic and forward-thinking community.

The chaos and experimentation in Kusama are complemented by its fast-paced development environment. Updates and changes are rolled out more frequently, enabling developers to quickly iterate on their projects and incorporate new features. This agility and responsiveness contribute to the rapid progress and evolution of the Kusama network.

Next up, we will cover Substrate, which forms the foundation of Polkadot and Kusama.

Introducing Substrate

Earlier, we delved into the worlds of Polkadot and Kusama. Now, it's time to explore the common foundation beneath both – Substrate. This versatile framework fuels the development of these blockchain ecosystems, offering a glimpse into the innovative potential it unlocks with its flexible and modular architecture, Substrate offers the following:

- **Modular components**: A wide array of modular building blocks, from consensus mechanisms to governance models

- **Tailored solutions**: Developers can configure components to craft highly customized and scalable blockchains

- **Abstraction of complexity**: Substrate abstracts low-level blockchain tasks, enabling developers to focus on unique features and smart contracts

- **Interoperability**: Substrate promotes collaboration by facilitating seamless communication between blockchains, fostering innovation

Substrate's robust architecture and comprehensive toolset make it a preferred framework for building powerful and customized dApps.

Substrate architecture

Substrate's architecture is distinguished by its flexibility and modularity. Central to this architecture is the **Runtime Module Library** (RML), a collection of reusable modules that empowers developers to craft tailored, feature-rich blockchains by selecting and configuring the modules that align with their project's needs. The following figure shows its architecture:

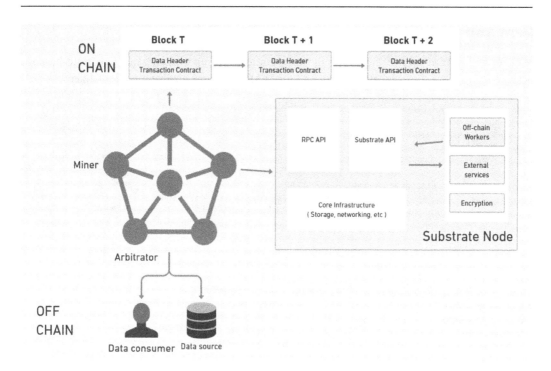

Figure 9.6 – Polkadot's Substrate architecture

The core of a Substrate-based blockchain is its runtime, which is responsible for executing the business logic and processing transactions. The runtime is customizable and can be tailored to meet the specific needs of the blockchain application. This allows developers to create efficient and specialized runtimes that optimize performance and resource utilization.

Substrate supports multiple consensus mechanisms, including **proof-of-stake (PoS)**, **proof-of-authority (PoA)**, and others. Developers can choose the consensus mechanism that best aligns with their blockchain's goals and requirements. This flexibility enables the creation of blockchains with different security models and performance characteristics.

Another key component of Substrate's architecture is the Substrate Node. The Substrate Node is responsible for maintaining the blockchain's state, validating transactions, and participating in consensus. It connects to the peer-to-peer network, allowing for communication and synchronization with other nodes in the network.

Now, let's delve into the core components that drive the operation and functionality of a Substrate-based blockchain: the client and runtime.

Client and runtime

The client and runtime play critical roles in the operation and functionality of a Substrate-based blockchain. *Figure 9.7* shows what the client and the WASM runtime environment look like:

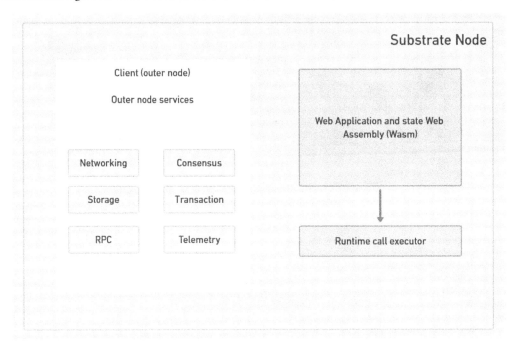

Figure 9.7 – The client and the WASM runtime environment

The client in Substrate is responsible for managing the interaction between users and the blockchain network. It provides the necessary interfaces and functionalities for users to connect to the network, submit transactions, and retrieve information from the blockchain. The client handles the communication with other nodes in the network and ensures the synchronization of the blockchain's state.

On the other hand, the runtime in Substrate contains the logic and rules that govern the behavior of the blockchain. It serves as the heart of the *Substrate-based blockchain*, executing smart contracts and enforcing the consensus rules. The runtime includes various modules that define the functionalities and features of the blockchain, such as token management, governance mechanisms, and custom business logic. Developers can customize the runtime by selecting and configuring the desired modules to create a blockchain tailored to their specific use case.

The separation of the client and runtime in Substrate allows for greater flexibility and modularity. Developers can choose different client implementations based on their specific needs, while still utilizing the same runtime logic. This separation also enables easier upgrades and maintenance of the client or runtime components independently.

Next, we'll look at various network types that Substrate supports.

Network types

Substrate supports a variety of network types, offering flexibility and adaptability to meet the diverse requirements of blockchain projects. These network types allow developers to tailor their Substrate-based blockchains for specific use cases, ranging from private networks to public networks with different levels of **permissioning**:

- **Private networks**: Private networks are restricted to a select group of participants, ensuring privacy and control over the network. These networks are suitable for consortiums, enterprises, or organizations that require a secure and private environment for internal operations. Private networks in Substrate can be set up with customized permissioning and access control.

- **Public networks**: Public networks are open to anyone who wants to participate, allowing for broader accessibility and decentralization. These networks promote inclusivity and encourage community engagement. Public networks built with Substrate can host various applications and facilitate the transfer of assets and information across a wide range of users.

- **Permissioned networks**: Permissioned networks strike a balance between private and public networks by allowing controlled access to specific participants. These networks are suitable for scenarios where certain participants need to be validated or authorized to join the network. Substrate enables the creation of permissioned networks with customizable access controls and governance models.

- **Hybrid networks**: Substrate also supports hybrid network configurations, combining elements of private and public networks. These networks provide the benefits of both private and public networks, allowing for a combination of privacy, control, and community engagement. Developers can configure hybrid networks based on their specific requirements and the desired balance between openness and privacy.

By offering various network types, Substrate provides developers with the flexibility to choose the appropriate network configuration for their specific use case. Whether it is a private network for internal operations, a public network for widespread participation, or a hybrid network that balances privacy and openness, Substrate enables the creation of tailored blockchains that align with specific needs and goals.

Now, let's look at various node types.

Node types

To ensure a consistent and current view of the blockchain state, network nodes are essential components of any blockchain system. These nodes synchronize with each other to maintain an updated copy of the blockchain and track incoming transactions. However, storing a complete copy of the entire blockchain can be resource-intensive in terms of storage and computing requirements. Additionally, downloading all the blocks from the genesis block to the latest can be impractical for many use cases.

To address these challenges and provide more efficient access to blockchain data while maintaining security and integrity, different types of nodes have been introduced. These nodes enable clients to interact with the blockchain in varying ways, based on their specific needs and resource constraints. These node types offer different levels of data storage and computational requirements, allowing clients to choose the most suitable option:

- **Full nodes**: Full nodes store a complete copy of the blockchain and participate in the network's consensus protocol. They require significant storage and computational resources but offer the highest level of security and independence, allowing for complete verification of the blockchain's history.

- **Light nodes**: Light nodes, also known as thin clients, do not store the entire blockchain. Instead, they rely on other full nodes for specific data retrieval, reducing storage and computational requirements. Light nodes sacrifice some independence and rely on trusted nodes for information.

- **Archive nodes**: Archive nodes specialize in maintaining a comprehensive historical record of the blockchain, storing all past transactions and states. They are valuable for research, auditing, and compliance purposes, but they require substantial storage capacity.

Let's embark on a deep dive into Substrate, uncovering its core components and intricate workings that make it a versatile framework for blockchain development.

Diving deep into Substrate

At this point, we have a basic understanding of the Substrate framework and can dive deeper by learning about some advanced concepts that will help us in the next chapter, when we have to use these concepts to build our blockchain.

In the realm of Substrate, runtime interfaces serve as vital contracts that ensure seamless compatibility and interoperability between the Substrate client and the blockchain runtime, especially during upgrades and cross-version interactions. We'll take a closer look in the next section.

Runtime interfaces

These interfaces specify the expected functions and data structures that facilitate interaction between the Substrate client and the blockchain.

For instance, a Substrate client can rely on the runtime interface to access APIs that are crucial for transaction validation, state management, and smart contract execution. When developers upgrade the blockchain's runtime, the defined runtime interface remains consistent, allowing for smooth and backward-compatible upgrades. This ensures that existing clients can continue to interact seamlessly with the blockchain, even in the presence of underlying changes and improvements.

Furthermore, runtime interfaces promote interoperability among different Substrate-based blockchains. By adhering to a shared runtime interface, Substrate-powered blockchains can communicate and collaborate, enabling cross-chain interactions. Developers can also leverage this flexibility to create custom runtime interfaces tailored to their specific use cases, enhancing the extensibility and adaptability of Substrate-based blockchains.

Now, let's venture into the foundational building blocks of Substrate. These are known as *core primitives*; they underpin the framework's essential functionalities and provide a framework for developing customized blockchains.

Core primitives

Core primitives form the building blocks of Substrate-based blockchains. These primitives provide essential functionality and structures for efficient data storage, event handling, and error management. Understanding these core primitives is crucial for developing robust and scalable blockchain solutions using Substrate.

Within the Substrate framework, core primitives serve as the bedrock upon which blockchain functionalities are built. These essential building blocks are as follows:

- **Storage**: Substrate offers a sophisticated storage system that enables efficient on-chain state management. Developers can structure and store data with precision, ensuring rapid and dependable access to the blockchain's state. This capability empowers the implementation of intricate data structures and the effective management of on-chain data:

  - *For example*, in Substrate, you can define and utilize storage items to store and retrieve data efficiently. For instance, to maintain a balanced ledger, you can use the following code snippet:

    ```
    decl_storage! {
        trait Store for Module<T: Trait> as Balances {
            Balances: map T::AccountId => Balance;
        }
    }
    ```

- **Events**: Events provide a mechanism to emit and monitor significant occurrences within the blockchain. Developers can craft custom events that represent specific blockchain actions or state changes. These events facilitate communication between the blockchain and external systems, enabling external applications to respond to blockchain events and initiate off-chain actions:

  - *For instance*, to define and use a custom event in Substrate, we can use the following code snippet:

    ```
    decl_event! {
        pub enum Event<T> where AccountId = <T as
    system::Trait>::AccountId {
            Transfer(AccountId, AccountId, Balance),
        }
    }
    ```

- **Error handling**: Robust error handling is fundamental in Substrate. The framework offers mechanisms for defining and managing errors that may arise during blockchain logic execution. Effective error management ensures the resilience and dependability of Substrate-based blockchains, allowing developers to address exceptional scenarios and communicate errors effectively:

  - To handle errors in Substrate, we can use custom error types and error propagation mechanisms in the blockchain logic.

  - These core primitives, when utilized in combination, empower developers to create efficient and reliable Substrate-based blockchains tailored to specific use cases and functionalities.

  - By leveraging these core primitives, developers can design efficient and scalable Substrate-based blockchains. The storage primitives enable organized and optimized data management, while events facilitate effective communication with external systems. Error handling mechanisms ensure the reliability and resilience of the blockchain's operations.

Next, we'll explore the dynamic world of FRAME.

FRAME

FRAME is a powerful framework within Substrate that provides a collection of reusable modules for building blockchain logic. FRAME simplifies the development process by offering pre-built, modular components that can be customized and combined to create customized Substrate-based blockchains.

Within Substrate's realm, **FRAME** emerges as a game-changing framework, offering a versatile toolkit of ready-to-use modules. These modules encompass a wide array of blockchain functionalities: balances, staking, identity, governance, treasury, and more. FRAME's modules are meticulously designed with modularity in mind, allowing developers to effortlessly incorporate or remove features according to their project's unique needs. A standardized structure and interface across modules ensure consistency and compatibility across diverse Substrate-based blockchains.

FRAME's true power lies in its adaptability. Developers can fashion custom modules by extending or adjusting existing ones, molding blockchain functionality to their specific use case. This flexibility empowers the creation of distinctive and purpose-built blockchains that align precisely with project requirements.

Additionally, FRAME champions compatibility and upgradability. Adhering to FRAME standards assures that a blockchain remains in harmony with future Substrate upgrades. This paves the way for the seamless integration of new features and bug fixes without compromising the blockchain's stability.

The following code snippet demonstrates how to include a staking module in a Substrate-based blockchain using FRAME:

```
use frame_system as system;
use pallet_staking as staking;
pub trait Config: system::Config + staking::Config {
```

```
    // Add any additional configuration types here
}
pub use pallet::*;
pub mod pallet {
    use frame_support::pallet_prelude::*;
    use frame_system::pallet_prelude::*;
    // Define the pallet struct
    pub struct Pallet<T>(_);
    // Implement pallet-specific functionality
    impl<T: Config> Pallet<T> {
        pub fn example_call(origin: OriginFor<T>) -> DispatchResult {
            // Custom blockchain logic goes here
            Ok(())
        }

        // Other functions can be added here
    }
```

This snippet demonstrates integrating the staking module into a Substrate-based blockchain using FRAME, allowing developers to customize staking functionality as needed.

Now, let's explore custom pallets, a vital concept within Substrate.

Building custom pallets

A pallet is a Substrate-specific term that refers to a module that encapsulates a specific set of functionalities within a Substrate-based blockchain.

Writing custom pallets allows developers to add custom functionalities and tailor their blockchain to specific use cases.

Custom pallets empower developers to extend the functionality of their Substrate-based blockchain beyond the standard modules offered by Substrate or FRAME. With custom pallets, developers can define and implement unique business logic, rules, and features tailored to their project's needs.

There are several benefits of using custom pallets:

- **Precision**: Custom pallets enable the creation of blockchain solutions precisely matched to the project's use case, delivering highly specialized functionality

- **Innovation**: Developers can stand out and innovate in a competitive landscape by offering unique features catering to specific industries or applications

- **Modularity**: Encapsulating related functionalities into pallets enhances code management and reusability, fostering collaboration within the Substrate community

- **Upgradability**: Pallets can be developed and upgraded independently, allowing seamless enhancements and bug fixes without disrupting the entire network

To create a custom pallet, you can follow these steps:

1. Define your pallet's purpose and functionalities.

2. Write the pallet's logic using Rust, adhering to Substrate's coding standards.

3. Integrate your custom pallet into your Substrate-based blockchain's runtime.

4. Test thoroughly to ensure functionality and compatibility.

5. Share your pallet with the community to foster collaboration and growth.

Now that we have discussed the important elements of Substrate, let's delve into a remarkable capability offered by Substrate – forkless runtime upgrades. These upgrades enable seamless protocol and runtime enhancements without the need for hard forks or node interruptions.

Forkless and runtime upgrades

One of the biggest features provided by Substrate is the ability to perform forkless runtime upgrades, in the sense you don't require forks for any software upgrade and the node doesn't have to be stopped for any update or upgrade – it can all be done at runtime. So, essentially, forkless upgrades refer to the ability to upgrade the blockchain's protocol and runtime without requiring a hard fork.

This means that the entire network does not need to undergo a disruptive split, thereby maintaining network continuity and preventing the creation of separate chains. Forkless upgrades are facilitated by Substrate's modular architecture and runtime interfaces. Developers can introduce changes to the protocol and runtime through runtime upgrades, ensuring that the network remains intact and smoothly transitions to the upgraded version.

Runtime upgrades allow developers to update the blockchain's runtime logic and functionality without interrupting the ongoing operations. With runtime upgrades, developers can introduce new features, fix bugs, or optimize performance without requiring a complete restart or disruption of the network. The upgraded runtime can be seamlessly applied to the existing blockchain, ensuring a smooth transition and preserving the state and history of the blockchain.

The benefits of forkless and runtime upgrades are significant. They provide a non-disruptive approach to evolving the blockchain, allowing for continuous improvements and enhancements while maintaining network consensus and preserving the integrity of the blockchain. Forkless upgrades and runtime upgrades ensure that Substrate-based blockchains can adapt to changing requirements, fix vulnerabilities, and introduce new features securely and efficiently.

By enabling forkless and runtime upgrades, Substrate empowers developers to create dynamic and upgradable blockchain networks that can evolve. It provides a robust mechanism for managing the life cycle of Substrate-based blockchains, allowing for seamless transitions and avoiding network disruptions that are typically associated with hard forks. Forkless and runtime upgrades contribute to the long-term sustainability and scalability of Substrate-based blockchains.

Another fundamental aspect that is shared across all the blockchain networks is the consensus mechanism. We'll discuss this next.

Consensus

Consensus in blockchain refers to the process by which participants agree on the state of the ledger. It ensures data integrity, security, and consistency among network nodes through various protocols. Consensus mechanisms are essential in determining the state of a blockchain, and this holds for all blockchain networks. With Substrate, a flexible framework for constructing blockchains, multiple consensus models are supported to enable nodes to reach an agreement. Each consensus model comes with its own set of advantages and trade-offs, making it crucial to carefully select the most suitable one for your chain. Substrate offers default consensus models that require minimal configuration, ensuring a straightforward setup process. However, if desired, it is also possible to develop a custom consensus model tailored to specific requirements. This flexibility empowers developers to tailor the consensus mechanism according to their unique needs, ensuring optimal performance and functionality for their blockchain network.

Let's explore the key consensus mechanisms that are used in various blockchain networks:

1. **Block authoring**: This phase involves nodes creating new blocks. During block authoring, nodes gather transactions, organize them into blocks, and propose them for inclusion in the blockchain. Block authoring is a pivotal process in the consensus mechanism. Block authoring refers to the creation and proposal of new blocks by participating nodes within the network. During the block authoring phase, nodes, also known as block authors or validators, collect and aggregate transactions from the network's mempool. They then organize these transactions into a block structure, appending them to the existing blockchain. Block authors play a critical role in ensuring the efficient and timely processing of transactions, as well as maintaining the overall integrity of the blockchain. Substrate provides a flexible environment for block authoring, allowing developers to design and implement their strategies based on their specific blockchain requirements. The framework offers a range of customizable features and modules that facilitate block authoring, such as transaction queues, consensus algorithms, and networking protocols. Substrate supports various block authoring mechanisms, including NPoS and Aura. NPoS is a widely used consensus algorithm in Substrate, where a set of validators is nominated based on stake ownership and selected to create blocks in a round-robin fashion. Aura, on the other hand, is a more deterministic block authoring mechanism that appoints a single validator to create blocks in each round.

2. **Block finalization**: This phase manages forks and determines the canonical chain. Block finalization ensures that conflicts or multiple versions of the blockchain are resolved, and a single, agreed-upon chain is selected as the authoritative version. Block finalization handles forks and determines the canonical chain within the network, ensuring consensus and maintaining the integrity of the blockchain. When multiple blocks are proposed during the block authoring phase, block finalization comes into play to select a single, agreed-upon chain as the canonical chain.

It resolves conflicts and establishes a definitive order of blocks, preventing diverging versions of the blockchain. Substrate employs various mechanisms for block finalization, including finality gadgets such as **GHOST-based Recursive Ancestor Deriving Prefix Agreement** (**GRANDPA**). This is a hybrid consensus algorithm that combines the benefits of PoS and finality gadgets. It utilizes a multi-round voting process where validators participate in a weighted voting system to determine the validity and order of blocks. Once a block receives sufficient votes and achieves supermajority approval, it becomes finalized, signifying its permanent position in the blockchain. This can be seen in the following figure:

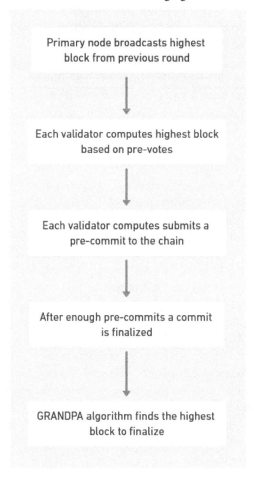

Figure 9.8 – How does the GRANDPA protocol work?

By separating consensus into these two phases, Substrate provides a modular approach to achieving agreement within the network. This division allows for flexibility in selecting consensus mechanisms that best suit the specific requirements and desired trade-offs of the blockchain. Developers using

Substrate can design their own block authoring and finalization mechanisms or choose from the existing options provided by the framework, tailoring the consensus process to their project's needs.

One widely recognized consensus model is **Proof of Work** (**PoW**), which is famously used in Bitcoin. In PoW, participants, known as miners, compete to solve complex mathematical puzzles. The first to solve the puzzle gets the right to validate transactions and add a new block to the blockchain. This process consumes significant computational power, making it computationally expensive and energy-intensive.

PoW's security lies in its difficulty – miners must expend real-world resources (electricity and computing power) to participate, deterring malicious actors. However, this energy consumption has raised environmental concerns.

While PoW offers robust security, it's not the most efficient solution. Blockchains such as Ethereum have transitioned to PoS, where validators are chosen to create new blocks based on the amount of cryptocurrency they *stake* as collateral. PoS aims to be more energy-efficient and is gaining popularity for its scalability and lower resource demands. Both PoW and PoS demonstrate how consensus mechanisms ensure network integrity and validity, albeit with differing trade-offs.

The finality that's achieved through block finalization ensures the immutability and consistency of the blockchain. It provides developers and users with confidence in the reliability of transactions and enables the construction of secure dApps.

With Substrate, developers have the flexibility to customize block finalization mechanisms to suit their specific blockchain requirements. They can integrate alternative finality gadgets or design finality algorithms, tailoring the consensus process to the unique needs of their blockchain network.

Let's explore the concept of deterministic finality, a crucial aspect that guarantees the irreversibility of transactions and offers a high degree of certainty regarding the blockchain's ultimate state.

Deterministic finality

Deterministic finality ensures the irreversibility of transactions and provides a high level of certainty regarding the final state of the blockchain.

In Substrate, deterministic finality guarantees that once a block is considered finalized, it becomes immutable and cannot be modified or reverted. This feature is achieved through the integration of finality gadgets, such as the GRANDPA consensus algorithm. GRANDPA utilizes a multi-round voting process among validators to reach an agreement on block finalization. Once a block achieves supermajority approval, it attains deterministic finality. This means that the transactions included in the finalized block are permanently recorded and considered the definitive state of the blockchain.

Deterministic finality provides several benefits. It enables developers and users to have confidence in the integrity of transactions as they are assured that finalized blocks cannot be tampered with. This reliability facilitates the building of robust dApps and supports secure economic interactions.

Summary

In summary, our exploration of the Polkadot ecosystem, Kusama, and Substrate has illuminated the realm of decentralized networks and blockchain development. We've delved into the innovative technologies, principles, and frameworks underpinning these interconnected platforms.

Polkadot's multi-chain architecture empowers scalability and interoperability in the blockchain space, enabling secure communication and data transfer across chains. This facilitates dApps and cross-chain collaborations, offering developers the opportunity to build specialized parachains tailored to unique use cases while benefiting from shared security and connectivity.

Kusama, the adventurous counterpart to Polkadot, provides an experimental playground for blockchain projects, fostering innovation and risk-taking in a fast-paced environment. Substrate, the versatile blockchain development framework, has played a pivotal role in both ecosystems, empowering developers to craft customized blockchain networks. Our journey has introduced us to essential concepts such as block authoring, finality, consensus mechanisms, and governance, all of which are essential for secure and reliable transactions within these networks. This newfound knowledge forms a solid foundation for further exploration and engagement in the dynamic world of blockchain technology.

As we conclude this chapter, our newfound knowledge of the Polkadot ecosystem, Kusama's experimental ethos, and Substrate's flexibility will serve as a solid foundation for further exploration and involvement in the exciting world of blockchain technology. In the next chapter, we will use the Substrate framework to create our own blockchain.

10

Hands-On with Substrate

By now, we have gained significant knowledge about blockchains and have even constructed our own custom blockchain using Rust. In this chapter, our objective is to utilize a popular blockchain framework called Substrate (which we learned about in the previous chapter, along with Polkadot and Kusama) to create our own blockchain. As we progress, you will witness how Substrate streamlines the process of building custom blockchains by managing many of the underlying complexities on our behalf. The aim of this chapter is to provide you with a comprehensive understanding of Substrate and equip you to harness its capabilities effectively.

Here's what we'll cover in this chapter:

- Building our own blockchain
- Simulating a network

Technical requirements

To continue our journey of building our own blockchain, we first need to have our environment set up. For detailed instructions on how to accomplish this, please refer to *Chapter 3*, *Building a Custom Blockchain*, where we provide step-by-step guidance for various operating systems, including installing Rust on Mac, Linux, and Windows.

Installing Substrate

With Rust installed and the Rust toolchains configured for Substrate development, you are now ready to complete the setup of your development environment. This involves cloning the Substrate node template files and compiling a Substrate node.

You can find detailed instructions for installing and setting up Substrate in the official Substrate documentation. Please refer to the following links based on your operating system:

- Linux development environment: `https://docs.substrate.io/install/linux/`

- MacOS development environment: `https://docs.substrate.io/install/macos/`

- Windows development environment: `https://docs.substrate.io/install/windows/`

With these steps completed, you're now equipped and ready to embark on your Substrate development journey. Happy coding!

Building our own blockchain

This section will guide you through building and initiating a functional single-node blockchain for practical exploration:

1. The first step in your journey as a blockchain developer is to understand the process of compiling and launching a single local blockchain node. We will use the Substrate framework. The Substrate node template provides a fully functional single-node blockchain that can be executed in your local development environment. It comes equipped with predefined components such as user accounts and account balances, enabling you to explore various common tasks.

2. By running the template without any modifications, you can have a functional node that generates blocks and supports transactions.

3. Once your local blockchain node is up and running, this section will demonstrate how to utilize a Substrate frontend template.

4. This will allow you to observe blockchain activities and submit transactions, providing practical insights into Substrate's functionalities as you begin your blockchain development journey.

Starting a local node

Let's get started with compiling and starting our own substrate node.

You first need to clone the node template repository by running the following command:

```
git clone https://github.com/substrate-developer-hub/substrate-node-template
```

By default, this command clones the main branch. However, if you have prior experience with Polkadot and wish to work with a specific version of it, you can use the `--branch` command-line option. This allows you to choose the particular branch that aligns with your requirements and development goals.

Change to the root of the node template directory by running the following command:

```
cd substrate-node-template
```

Create a new branch to contain your work:

```
git switch -c my-learning-branch-2023-12-11
```

Compile the node template by running the following command:

```
cargo build –release
```

When building optimized artifacts, it is recommended to consistently include the --release flag. On the initial compilation, the process may take some time to complete.

Upon successful completion, you should see a line similar to this:

```
Finished release [optimized] target(s) in 11m 23s
```

Once your node has finished compiling, you can begin exploring its functionalities with the frontend template that we will download, but before that, let's first start the node.

In the same terminal where you compiled your node, you can now start the node in development mode by executing the following command:

```
./target/release/node-template --dev
```

The node-template command-line options determine the node's behavior during runtime. In this context, the --dev option indicates that the node will run in development mode, utilizing the predefined development chain specification. Additionally, when the node is stopped (e.g., by pressing *Ctrl + C*), the --dev option automatically deletes all active data, including keys, the blockchain database, and networking information. As a result, using the --dev option guarantees a clean working state each time you stop and restart the node, allowing for a fresh and consistent environment.

To ensure your node is successfully up and running, review the output displayed in the terminal.

The terminal should exhibit output similar to this:

```
2023-12-11 09:21:34 Substrate Node

2023-12-11 09:21:34 ✋  version 4.0.0-dev-de262935ede

2023-12-11 09:21:34 ♥  by Substrate DevHub <https://github.com/
substrate-developer-hub>, 2017-2022

2023-12-11 09:21:34 📋  Chain specification: Development

2023-12-11 09:21:34 🏷  Node name: limping-oatmeal-7460
```

```
2023-12-11 09:21:34 ♟ Role: AUTHORITY

2023-12-11 09:21:34 ▤ Database: RocksDb at /var/folders/2_/
g86ns85j517fdn1621ptzn500000gn/T/substrate95LPvM/chains/dev/db/full

2023-12-11 09:21:34 ▦  Native runtime: node-template-100 (node-
template-1.tx1.au1)

2023-12-11 09:21:34 ⚒ Initializing Genesis block/state (state:
0xf6f5…423f, header-hash: 0xc665…cf6a)

2023-12-11 09:21:34 ☺ Loading GRANDPA authority set from genesis on
what appears to be first startup.

2023-12-11 09:21:35 Using default protocol ID "sup" because none is
configured in the chain specs

2023-12-11 09:21:35 🏷  Local node identity is:
12D3KooWQELo5BwrmYcG6sb5zrj3t2pSyY3QRJVDVqRFxy8NUkGX


    . . .


    . . .


    . . .


    . . .


2023-12-11 09:34:02 ⌗ Idle (0 peers), best: #3 (0xcdac…26e5),
finalized #1 (0x107c…9bae), ↓ 0 ↑ 0
```

As you monitor the log, notice the incrementing number after finalized using the instruction on the repo at https://github.com/nvm-sh/nvm. This signifies the active generation of new blocks and the consensus achieved regarding the represented state within your blockchain.

In future sections, we'll delve deeper into analyzing the log output. But for now, you can comprehend that your node is operational, successfully creating blocks.

To continue, ensure the terminal displaying the node output remains open.

Installing a frontend template

To engage with our blockchain node, Substrate provides a ready-to-use frontend template. Let's install it and explore its functionality.

Firstly, it's essential to confirm the presence of Node.js on your system, ensuring it meets the required version specifications, such as v19.7.0. Open a terminal and check the Node.js version using the following command:

```
node --version
```

If the command doesn't display a version number, it indicates that Node.js isn't installed on your system. In such a case, you can install it using nvm.

Remember, Node.js needs to be at least version v14 for running the frontend template.

Additionally, ensure you have Yarn installed for further setup:

```
yarn -version
```

To execute the frontend template, ensure that your Yarn version is at least v3. If you have an older version, updating can be done by specifying a version number through the yarn version command. In case this command doesn't yield a version number or you require assistance installing a specific Yarn version, refer to the installation guidelines available on the Yarn website (https://classic.yarnpkg.com/lang/en/docs/install/).

Now, proceed to clone the repository containing the frontend template using this command:

```
git clone https://github.com/substrate-developer-hub/substrate-front-
end-template
```

Now, let's change to the root of this directory that we have cloned:

```
cd substrate-front-end-template
```

Now, it's crucial to install all the dependencies listed in the Yarn files into your system. When you clone a Node.js or Yarn project, you're importing specification files of the necessary dependencies to run the project. However, you won't import the actual required dependencies themselves as they tend to be large in size. This step involves separately installing these dependencies:

```
yarn install
```

Starting the frontend template

The Substrate frontend template is equipped with user interface components that facilitate interaction with the Substrate node, allowing you to perform various common tasks.

To utilize the frontend template, follow these steps:

1. Ensure that your current working directory corresponds to the root directory where you installed the frontend template in the previous section.

2. Initiate the frontend template by executing the following command:

```
yarn start
```

3. Normally, this command automatically opens `http://localhost:8000` in your default browser. If required, you can manually enter the URL (`http://localhost:8000`) to access the frontend template.

4. The upper section of the template includes an **Account selection** list, enabling you to choose the account you wish to work with when performing on-chain operations. Additionally, this section provides information about the connected chain:

Figure 10.1 – Blockchain dashboard

5. You may also observe a **Balances** table in the frontend template, which contains a set of predefined accounts, some of which come with preconfigured funds. This sample data allows you to experiment with operations such as fund transfers:

Balances

Figure 10.2 – Balances table

Transferring the funds

Now that you have a blockchain node up and running on your local computer and access to the frontend template for conducting on-chain actions, you are all set to explore various ways to interact with the blockchain.

The default frontend template comes equipped with numerous components that enable you to try out different common tasks. For this section, let's start by performing a straightforward fund transfer operation, where funds will be moved from one account to another.

To transfer funds to an account, follow these steps:

1. In the **Balances** table, observe the predefined accounts, such as **Max**, which currently have no funds associated with them:

Balances

Name	Address		Balance
Eve	5GrwvaEF5zXb26Fz9rcQpDWS57CtERHpNehXCPcNoHGKutQY		1.1529 MUnit
Eve_stash	5GNJqTPyNqANBkUVMN1LPPrxXnFouWXoe2wNSmmEoLctxIZY		1.1529 MUnit
John	5FHneW46xGXgs5mUiveU4sbTyGBzmstUspZC92UhjJM694ty		1.1529 MUnit
John_stash	5HpG9w8EBLe5XCrbczpwq5TSXvedjrBGCwqxK1iQ7qUsSWFc		1.1529 MUnit
charlie	5FLSigC9HGRKVhB9FiEo4Y3koPsNmBmLJbpXg2mp1hXcS59Y		0
Max	5DAAnrj7VHTznn2AWBemMuyBwZWs6FNFjdyVXUeYum3PTXFy		0
Joe	5HGjWAeFDfFCWPsjFQdVV2Msvz2XtMktvgocEZcCj68kUMaw		0
ferdie	5CiPPseXPECbkjWCa6MnjNokrgYjMqmKndv2rSnekmSK2DjL		0

Figure 10.3 – Predefined user's balance in the table

2. Beneath the **Balances** table, you'll find the **Transfer** component. Use this component to initiate the fund transfer from one account to another.

3. Choose **Max** from the list of available accounts to populate the recipient address for the fund transfer.

4. Specify an amount of at least 1000000000000 as the sum to be transferred, and then click the **Submit** button:

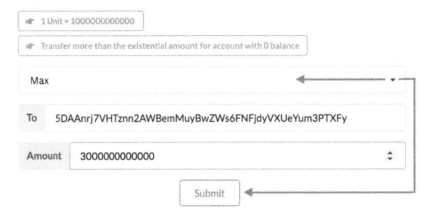

Figure 10.4 – Dashboard to transfer the fund

5.   Observe that the values in the **Balances** table get updated, reflecting the successful transfer:

Figure 10.5 – Updated balances table

To view events related to the transfer you just made, check the **Events** component.

The Substrate blockchain communicates the results of asynchronous operations through events. By using the **Events** component, you can examine details about each operation conducted during the transfer:

Figure 10.6 – Event details

For instance, once the transaction is completed and included in a block, you will see a confirmation message similar to the following:

```
☺ Finalized. Block hash:
0xa50d3f99fcea8a1611806895aa3f4d4d55fdc4989fbb2148d4856a043d01f808
```

Next up, we will see how we can simulate a network.

Simulating a network

In this section, we will extend our knowledge from single-node operations to multi-node setups. We'll focus on setting up a private blockchain network using an authority set of private validators. Within this network, Substrate employs an authority consensus model, where a predefined list of authorized accounts, known as authorities, takes turns in a round-robin fashion to create blocks. This fundamental introduction will equip you with the skills to work with private blockchain networks and understand authority-based consensus mechanisms. Throughout this chapter, you will experience the practical functioning of the authority consensus model by utilizing two preconfigured accounts as the authorities responsible for block production. This simulated network will involve running the two nodes on a single computer, each with different accounts and keys. In the upcoming section, we will begin our journey by initiating a blockchain node using a predefined account, allowing us to explore Substrate's functionality further.

Starting the first blockchain node

Before generating keys to set up your private Substrate network, you can grasp the fundamental principles using a predefined network specification known as `local`. This specification runs with preconfigured user accounts, allowing you to learn the basics without generating new keys.

We will now simulate a private network by running two Substrate nodes on a single local computer. These nodes will utilize predefined accounts named `Eve` and `Charlie`.

To initiate the blockchain, follow these steps:

1. Open a terminal shell on your computer.

2. Navigate to the root directory where you compiled the Substrate node template.

3. Purge old chain data by executing the following command:

    ```
    ./target/release/node-template purge-chain --base-path /tmp/Eve
    --chain local
    ```

    Alternatively, execute this command:

    ```
    cargo run --release -- purge-chain --base-path /tmp/Eve --chain
    local
    ```

4. The command will prompt you to confirm the operation:

    ```
    Are you sure to remove "/tmp/Eve/chains/local_testnet/db"?
    [y/N]:
    ```

5. Type y to confirm that you want to remove the chain data. Remember, it's essential to remove old chain data when starting a new network.

6. Start the local blockchain node using the Eve account by executing the following command as a single command:

    ```
    ./target/release/node-template \
    ```

    Alternatively, execute this command:

    ```
    cargo run --release --
    --base-path /tmp/Eve\
    --chain local \
    --Eve- \
    --port 30333 \
    --ws-port 9945 \
    --rpc-port 9933 \
    --node-key 000000000000000000000000000000000000000000000000000000000000000
    00000000001 \
    --telemetry-url "wss://telemetry.polkadot.io/submit/ 0" \
    --validator
    ```

7. Once you put in the previous command, the node should start, and if it does, you will receive a message like this in the terminal:

    ```
    2023-12-13 15:29:55 Substrate Node

    2023-12-13 15:29:55 ✌  version 4.0.0-dev-de262935ede

    2023-12-13 15:29:55 ♥  by Substrate DevHub <https://github.com/
    substrate-developer-hub>, 2017-2022

    2023-12-13 15:29:55 📑 Chain specification: Local Testnet

    2023-12-13 15:29:55 🏷  Node name: Eve

    2023-12-13 15:29:55 👤 Role: AUTHORITY

    2023-12-13 15:29:55 💾 Database: RocksDb at /tmp/eve/chains/
    local_testnet/db/full

    2023-12-13 15:29:55 🖳  Native runtime: node-template-100 (node-
    template-1.tx1.au1)

    2023-12-13 15:29:55 🔨 Initializing Genesis block/state (state:
    0x6894…033d, header-hash: 0x2cdc…a07f)
    ```

```
2023-12-13 15:29:55 ⊚ Loading GRANDPA authority set from
genesis on what appears to be first startup.

2023-12-13 15:29:56 Using default protocol ID "sup" because none
is configured in the chain specs

2023-12-13 15:29:56 ⟨⟩  Local node identity is:
12D3KooWEyoppNCUx8Yx66oV9fJnriXwCcXwDDUA2kj6vnc6iDEp

2023-12-13 15:29:56 ▤ Operating system: macos

2023-12-13 15:29:56 ▤ CPU architecture: x86_64

2023-12-13 15:29:56 ⟨⟩ Highest known block at #0

2023-12-13 15:29:56 ⋏ Prometheus exporter started at
127.0.0.1:9615

2023-12-13 15:29:56 Running JSON-RPC HTTP server:
addr=127.0.0.1:9933, allowed origins=Some(["http://
localhost:*", "http://127.0.0.1:*", "https://localhost:*",
"https://127.0.0.1:*", "https://polkadot.js.org"])

2023-12-13 15:29:56 Running JSON-RPC WS server:
addr=127.0.0.1:9945, allowed origins=Some(["http://
localhost:*", "http://127.0.0.1:*", "https://localhost:*",
"https://127.0.0.1:*", "https://polkadot.js.org"])

2023-12-13 15:29:56 creating instance on iface 192.168.1.125

2023-12-13 15:30:01 ⊞ Idle (0 peers), best: #0 (0x2cdc…a07f),
finalized #0 (0x2cdc…a07f), ↓ 0 ↑ 0
```

Pay special attention to the following messages displayed in the output:

- ⚒ Initializing Genesis block/state (state: 0x6894…033d, header-hash: 0x2cdc…a07f): This message identifies the initial or genesis block that the node is currently utilizing. When you launch the next node, ensure that these values are the same for proper network synchronization.

- ⟨⟩ Local node identity is: 12D3KooWEyoppNCUx8Yx66oV9fJnriXwCcXwDDUA2kj6vnc6iDEp: This message indicates a unique string that identifies this specific node. The string is determined by the --node-key used when starting the node with the Eve account. You will use this string to identify and connect to this node when setting up a second node.

- 2022-08-16 15:30:01 ⊞ Idle (0 peers), best: #0 (0x2cdc…a07f), finalized #0 (0x2cdc…a07f), ↓ 0 ↑ 0: This message indicates that there are currently no other nodes in the network, and no blocks are being produced. It suggests that another node needs to join the network before block production can commence.

Next, we'll expand our network by adding another node, this time using the Charlie account keys. This process will allow us to connect multiple nodes and further explore Substrate's capabilities.

Adding more nodes

Now that the node initiated with the Eve account keys is up and running, you can proceed to add another node to the network using the Charlie account. As you are joining an already active network, the running node will serve as an identifier for the new node to connect with. The commands to accomplish this are quite similar to the ones used previously in the *Starting the first blockchain node* section, but with some crucial differences.

To add a node to the running blockchain, follow these steps:

1. Open a new terminal shell on your computer.

2. Navigate to the root directory where you compiled the Substrate node template.

3. Purge old chain data by executing the following command:

    ```
    ./target/release/node-template purge-chain --base-path /tmp/
    Charlie --chain local -y
    ```

4. By including the y flag in the command, you can bypass the prompt for confirmation while removing chain data.

5. Start a second local blockchain node using the Charlie account with the following command:

    ```
    ./target/release/node-template database new --db-path /path/to/
    Charlie_database --db-instance charlie_instance

    ./target/release/node-template \
    --base-path /tmp/Charlie \
    --chain local \
    --Charlie \
    --port 30334 \
    --ws-port 9946 \
    --rpc-port 9934 \
    --telemetry-url "wss://telemetry.polkadot.io/submit/0" \
    --validator \
    --bootnodes/ip4/127.0.0.1/tcp/30333/
    p2p/12D3KooWEyoppNCUx8Yx66oV9fJnriXwCcXwDDUA2kj6vnc6iDEp \
    --database-instance charlie_instance
    ```

Take note of the following differences between this command and the previous one. As the two nodes are running on the same computer, you must specify distinct values for the --base-path, --port, --ws-port, and --rpc-port options.

- This command includes the bootnodes option and specifies a single boot node, which is the node initiated by Eve.

- The --bootnodes option provides the following information:

  - ip4 indicates that the IP address for the node is in IPv4 format.

  - 127.0.0.1 specifies the IP address for the running node, which, in this case, is the localhost.

  - tcp designates TCP as the protocol used for peer-to-peer communication.

  - 30333 denotes the port number used for peer-to-peer communication, specifically the port number for TCP traffic.

  - 12D3KooWEyoppNCUx8Yx66oV9fJnriXwCcXwDDUA2kj6vnc6iDEp identifies the running node that this new node will communicate with within the network. In this case, it refers to the node initiated using the Eve account.

We'll verify that our blockchain network is functioning correctly by confirming block production and examining essential information about the blockchain.

Verifying block production

Once you start the second node, both nodes should establish a peer-to-peer connection with each other and the process of producing blocks will commence.

To confirm that blocks are being successfully finalized, check the terminal where you started the first node. Look for lines similar to the following:

```
2023-12-13 15:32:33 discovered:
12D3KooWBCbmQovz78Hq7MzPxdx9d1gZzXMsn6HtWj29bW51YUKB /ip4/127.0.0.1/
tcp/30334

2023-12-13 15:32:33 discovered:
12D3KooWBCbmQovz78Hq7MzPxdx9d1gZzXMsn6HtWj29bW51YUKB /ip6/::1/
tcp/30334

2023-12-13 15:32:36 🦀 Starting consensus session on top of parent
0x2cdce15d3...

2023-12-13 15:32:36 🎁 Prepared block for proposing at 1 (5 ms)
[hash: 0x9ab34...4104b; parent_hash: 0x2cdc...a07f; extrinsics (1):
[0x4634...cebf]]
```

```
2023-12-13 15:32:36 🗝 Pre-sealed block for proposal at 1. Hash now
0xf0869...9847, previously 0x9ab34...4104b.

2023-12-13 15:32:36 ⛄ Imported #1 (0xf086…9847)

2023-12-13 15:32:36 🖧 Idle (1 peers), best: #1 (0xf086…9847),
finalized #0 (0x2cdc…a07f), ↓ 1.0kiB/s ↑ 1.0kiB/s


...


2023-12-13 15:32:48 👥 Starting consensus session on top of parent
0x0d5ef319...

2023-12-13 15:32:48 🎁 Prepared block for proposing at 3 (0 ms)
[hash: 0xa307c...601346a; parent_hash: 0x0d5e...2a7f; extrinsics (1):
[0x63cc...39a6]]

2023-12-13 15:32:48 🗝 Pre-sealed block for proposal at 3. Hash now
0x0c556...cfb3d51b, previously 0xa307c...601346a.

2023-12-13 15:32:48 ⛄ Imported #3 (0x0c55…d51b)

2023-12-13 15:32:51 🖧 Idle (1 peers), best: #3 (0x0c55…d51b),
finalized #1 (0xf086…9847), ↓ 0.7kiB/s ↑ 0.9kiB/s
```

In these lines, you can find essential information about your blockchain:

- The second node's identity is discovered on the network (12D3KooWBCbmQovz78Hq7MzPxdx9d1gZzXMsn6HtWj29bW51YUKB)

- The node has one peer (1 peer)

- The nodes have successfully produced some blocks (best: #3 (0x0c55…d51b))

- The blocks are being finalized (finalized #1 (0xf086…9847))

Make sure you observe similar output in the terminal where you initiated the second node.

To shut down one of the nodes, press *Ctrl + C* in the terminal shell.

After you shut down the node, you'll notice that the remaining node has zero peers and has stopped producing blocks, as shown in the following example:

```
2023-12-13 15:53:45 🖧 Idle (1 peers), best: #143 (0x8f11…1684),
finalized #141 (0x5fe3…5a25), ↓ 0.8kiB/s ↑ 0.7kiB/s
2023-12-13 15:53:50 🖧 Idle (0 peers), best: #143 (0x8f11…1684),
finalized #141 (0x5fe3…5a25), ↓ 83 B/s ↑ 83 B/s
```

Shut down the second node as well by pressing *Ctrl + C* in the terminal shell .

If you wish to remove the chain state from the simulated network, employ the `purge-chain` subcommand along with the `--base-path` command-line options for the `/tmp/Charlie` and `/tmp/Eve` directories.

Summary

In this chapter, we embarked on the journey of building our very own blockchain using Substrate. Through hands-on experience and practical exercises, we gained a comprehensive understanding of setting up a Substrate node on our local machine and simulating a network of multiple nodes. We explored the inner workings of blockchain technology, witnessing how block production is limited to a rotating list of authorized accounts, known as *authorities*, who take turns creating blocks.

Adding a second node to our blockchain allowed us to witness decentralized node interactions and block production in a live environment. Throughout, we learned about essential operations, such as purging chain data and managing node identities, as well as command-line options for configuring nodes. This hands-on experience and practical knowledge will equip us for further blockchain development endeavors using Substrate.

In the next chapter, we will delve into the future of Rust for blockchains. After our extensive journey into blockchain development, it's crucial to explore the promising prospects Rust offers for blockchain technology. We'll discuss Rust's role in various blockchains and Web3 technologies, the growing Rust community's interest in blockchain development, and career opportunities in the Web3 space. Additionally, we'll explore potential projects and how to extend our knowledge beyond this book, providing valuable insights for readers looking to advance their careers in blockchain development.

Part 5:
The Future of Blockchains

In the final part of this book, we will explore what the future of blockchains looks like when it comes to Rust.

This part has the following chapter:

- *Chapter 11, Future of Rust for Blockchains*

Future of Rust for Blockchains

We've been together on this long journey of learning about blockchains and even building blockchains from scratch, and we can say that we have a great amount of information and some experience working with blockchains.

It might now be clear to you that blockchains have a great future because of all the benefits they provide – immutability, decentralization, and trust. As we know, mass adoption has even started in some industries and it's only a matter of time before blockchains become mainstream.

In this chapter, we will cover the following topics:

- What the future looks like for Rust blockchains
- Upcoming blockchains
- Upcoming Rust Web3 projects

What the future looks like for Rust blockchains

In this section, we will go through all the popular and upcoming blockchains that are using Rust as the primary technology and learn what the future looks like for these blockchains. This will help you understand which particular blockchain you can focus your efforts on and which community would be the most beneficial for you to join. Some of the popular blockchains can be seen in the following figure:

Figure 11.1 – Popular and upcoming blockchains

Popular blockchains

Let's first start by talking about the blockchains that are already well established and have a great number of users, dApps, and active communities.

Solana

The growth story of **Solana** has been nothing short of phenomenal; according to `https://solanaproject.com/`, there are a total of 394 projects building on the Solana blockchain with a total locked value of more than 600 million dollars. Solana is able to support more than 4,000 **transactions per second** (TPS).

This rapid adoption of Solana can be attributed to its high throughput, low transaction fees, and robust developer tools, which have attracted a diverse range of projects across various industries.

Solana has surpassed two million total users with the continuous development and expansion of the platform. The community surrounding Solana has also been growing steadily, with active participants ranging from developers and investors to enthusiasts and supporters.

Solana, with its ever-improving features and a growing developer community, is poised to attract a larger audience of developers and users. The platform's commitment to nurturing a developer-friendly ecosystem fosters innovation and diversifies the range of applications built on Solana. Moreover, as the blockchain industry gains mainstream recognition and adoption, Solana's efficient performance and scalability make it an attractive choice for projects seeking robust and reliable blockchain infrastructure.

Rust has also proven to be highly effective in the Solana blockchain for several reasons:

- **Emphasis on safety and memory management**: Rust's focus on safety and efficient memory management is crucial for building a secure and dependable blockchain infrastructure
- **Efficiency and concurrency**: Solana's unique Proof of History consensus mechanism demands high efficiency and concurrency—areas where Rust excels
- **Low-level programming**: Rust's ability to handle low-level programming with minimal overhead aligns perfectly with Solana's requirements
- **Strict type system**: Rust's strict type system enhances the protocol's stability
- **Robust error handling**: Rust's robust error handling contributes to the resilience of the Solana protocol

For a more detailed exploration of these advantages, refer to *Chapter 7, Exploring Solana by Building a dApp*, where we delve into Rust's significance in blockchain development, specifically in the context of Solana.

NEAR

According to **NEAR's** official website (`https://near.org/`), in terms of sheer numbers, NEAR is highly impressive – more than 300,000 daily transactions, more than 20,000 new accounts created on a daily basis, and an average of 50,000 daily active accounts. NEAR is impressive for many more reasons:

- NEAR's growth prospects are optimistic, with ongoing enhancements and developer tools attracting more dApps and users
- NEAR's commitment to user-friendly experiences and scalability positions it well for adoption in DeFi, NFTs, and gaming
- Active community and partnerships drive innovation and collaboration in the NEAR ecosystem
- Rust's concurrency support is essential for processing multiple transactions and smart contracts simultaneously
- Rust's safety features minimize bugs and vulnerabilities in NEAR's smart contracts, enhancing asset security
- Rust's cross-compilation capabilities ensure efficient operation on various devices and operating systems
- The extensive Rust community and ecosystem empower NEAR developers to build and maintain robust applications

You can refer to *Chapter 8* for more details on NEAR's blockchain and Rust usage.

Polkadot

According to the official **Polkadot** website (`https://www.polkadot.network/`), in the Polkadot ecosystem, there are already more than 90 parachains, even though there can only be a total of 100 parachains. Since there are multiple parachains, the number of dApps has reached 300, which is quite high considering parachains went live just two years ago.

More than 550 projects are actively building on Polkadot and there are 2,500+ monthly active users; this clearly shows that Polkadot has a great future.

Apart from Polkadot, Kusama and Substrate also have a great number of users and are being adopted heavily. As we already know, both networks Polkadot and Kusama are built with Rust and Substrate, but the framework used to build these is also built with Rust. What's more interesting is that Ink! – the language used to build smart contracts on Polkadot – is also based on Rust.

Since Polkadot is a highly complex technology, Rust was the right choice due to the speed, weight, and features it provides.

Rust has played a crucial role in enabling Polkadot to scale effectively. As a blockchain platform designed for interoperability and scalability, Polkadot requires a programming language that can handle complex and performance-critical tasks. Rust's unique features have made it an excellent fit for the requirements of building a scalable and secure blockchain network such as Polkadot.

Aptos

With 110 active validators, more than 200 million total transactions, 890 million total staked tokens, and more than 200 dApp projects deployed, with some of them (such as Liquidswap) already having more than 160 K unique users, **Aptos** is a blockchain that's growing extremely fast. We can learn more about this from the official website for Aptos at `https://aptosfoundation.org/`.

Based on the Diem blockchain project by meta, Aptos uses *Move* as the primary programming language for its smart contracts, where Move is derived from Rust, making it highly performant as well as stable and scalable. More information is available at `https://github.com/move-language/move`. We will explore Diem blockchain in greater detail in the next section.

Aptos uses **proof-of-stake** (**PoS**) with *BFT consensus* and *staked validators* that are segmented into full nodes and light clients and follow a modular design where changes are introduced to individual nodes.

Aptos also has a parallel transaction engine with BlocksSTM to increase throughput and logical data models where resources cannot be discarded. This enables the concurrent execution of all key stages of the transaction and introduces a structured path to add scaling validators, leading to unified state-sharing to boost validators' performance.

This means that Aptos is able to provide extremely high TPS (around 160,000), which is one of the highest in the industry, and is also highly scalable. Building a highly scalable and fast blockchain has really helped Aptos, since they were able to raise $200 million, meaning they will be able to fund a lot of projects with grants. This means more projects will build on Aptos or even shift to Aptos.

Chingaari, an app with 90 million users, recently shifted to Aptos from Solana owing to the downtimes that Solana faces from time to time, which have been heavily criticized, and owing to the speed and scalability that Aptos provides.

Sui

According to Sui Scan, the explorer for the **Sui** blockchain, the total number of transactions has already exceeded 800 million, with a recent surge in activity due to the Sui 8192 game that was recently released, raising the market cap to $461 million, which is huge.

While the total number of projects on Sui does not exceed 100 as of now, the fact that this is also a Move blockchain and a direct competitor of Aptos means that many developers will favor Sui as an alternative to Aptos. As ecosystems saturate, developers often opt for similar environments to capitalize on reduced competition, facilitating growth and development.

Sui is more focused on asset ownership and dynamic assets and claims to have security built in, which simply means secure assets, secure contracts, and secure transactions, leading to an overall secure network.

All in all, Sui is on a hyper-growth trajectory and the best part is that it's built to scale, so it will see a lot of traction and multiple projects adopting it quickly. Aptos and Sui look like the future blockchains that will dominate the industry, and it's great that they're based on Rust.

In this section, we have learned about how popular Rust blockchains are growing and we have also discussed their future. These blockchains already have quite an edge over their competitors due to the fact that they've adopted Rust, which is lightweight, fast, and feature-rich as we've already seen in this book. This makes the future of these blockchains very optimistic.

Let's now talk about some of the new blockchains that are slowly gaining popularity.

Upcoming blockchains

These are the blockchains that need to be on your radar; they are growing fast and getting a lot of traction. It can be highly beneficial to be a part of the communities of these blockchains because there's way less competition, you don't have to struggle for resources, and grants get approved much easier. You may get greater support from the community because there's a lot of push from the blockchain for builders to get started and build. As more projects get built on the chain, the more users there will eventually be for the chain.

Diem

Diem (`https://www.diem.com/en-us/`) is a Rust-based project that has encountered numerous challenges throughout its history. While the future of Diem may seem uncertain as we explore upcoming blockchains, it remains important to discuss this blockchain for a couple of key reasons. Firstly, Diem played a significant role in the innovation of the Move programming language. Secondly, it is possible that Diem could attract further investments in the future, making it recommended for developers to be aware of its existence. Diem (formerly known as *Libra*) is a blockchain-based project initiated by Facebook (now Meta Platforms, Inc.) with the goal of creating a global digital currency and financial infrastructure. The project was announced in June 2019 but faced significant regulatory scrutiny and challenges, which led to its development being delayed and rebranded as Diem.

Diem aimed to create a new global digital currency called the **diem**. This digital currency was intended to be a stablecoin, meaning its value would be pegged to a basket of stable assets such as major fiat currencies, with the aim of reducing the price volatility often associated with cryptocurrencies such as Bitcoin.

Diem's distinctive features lie in its ambition to establish a global digital currency and financial infrastructure, mitigating price volatility through stablecoin technology. Its initial consortium, the Libra Association, aimed for a unified currency, later pivoting to individual fiat-linked stablecoins in response to regulatory concerns, showcasing adaptability and innovation. The Diem project was initially governed by the Libra Association, a consortium of various companies and organizations, with Facebook (Meta) being one of the founding members. However, due to regulatory concerns and pressure, several prominent members, including PayPal, Mastercard, Visa, and others, withdrew their support from the project.

The announcement of Diem (Libra) raised concerns and regulatory questions from governments around the world, particularly related to issues such as financial stability, money laundering, consumer protection, and privacy. This led to intense scrutiny and delays in the project's development.

In December 2020, the project was rebranded from Libra to Diem as a part of an effort to address regulatory concerns and to create a more distinct identity. The project also shifted its approach by focusing on creating stablecoins tied to individual fiat currencies rather than on a single global digital currency.

Even though the project is keeping a low profile as of now, there is a high chance of the project being renewed.

Massa

Massa (`https://massa.net/`) is an emerging project with a unique selling point: it boasts low transaction fees for both transactions and smart contracts. This cost-effectiveness is attributed to the blockchain's minimal infrastructure requirements, which result in reduced operating expenses for node operators. This, in turn, contributes to the blockchain's low transaction fees. Furthermore, Massa emphasizes the balanced distribution of tokens and infrastructure, aiming to establish itself as a genuinely decentralized blockchain platform. This blockchain is PoS and hence consumes less energy.

Massa employs a *multithreaded block graph* that allows multiple threads (parallel processes) to work together to validate and process transactions simultaneously. This innovative approach significantly enhances the blockchain's capacity to generate blocks concurrently, resulting in a remarkable throughput of 10,000 transactions per second.

Through its parallel block framework, Massa achieves a groundbreaking feat as the inaugural blockchain technology in addressing the blockchain trilemma – the challenging balance between decentralization, security, and scalability.

This is a new generation Rust blockchain where new generation blockchains are trying to make transactions cheaper and faster. The vision is to reach *Web2* speeds so that blockchains blend in behind traditional payment systems and are adopted as the new way of financial transactions.

Imagine a future where all financial transactions occur solely on blockchains. In just a few years, these emerging blockchains could reap substantial benefits from this vision. While it might seem distant, we are rapidly progressing towards this new reality.

Phala

Phala Network (`https://phala.network/`) is spearheading a transformation in the realm of *Web3* by offering dApp developers an off-chain computing infrastructure that is genuinely decentralized and devoid of trust dependencies. Through the linkage of Smart Contracts to our off-chain entities, referred to as **Phat Contracts** (`https://phala.network/phat-contract`), developers can infuse their dApps with a dynamic blend of cross-chain integrations, internet connectivity, and robust computational capabilities. Phat Contracts elevate the intelligence of Smart Contracts, and seamlessly integrating them can be accomplished in mere minutes using the user-friendly *Phat Bricks*, which require no coding expertise.

The landscape of Web3 development consistently encounters the technical confines of **on-chain construction**. Contemporary dApps demand more than just Smart Contracts to accommodate feature-rich functionalities. As Web3 has matured, it has become evident that efficient **off-chain computation**, which involves performing complex calculations and tasks outside the blockchain network, will be indispensable for numerous standard dApp use cases. Phala empowers dApp developers with access to potent off-chain services while upholding the core tenets of Web3. This signifies the true essence of computation in its optimal form. Moreover, in an increasingly centralized digital environment, Phala Network is charting a divergent course by erecting a computational network that is open to anyone for contribution or establishing atop, all within the realm of trustlessness. To put it succinctly, Phala embodies a Public Goods Network.

Phala Network is meticulously crafted to offer a multi-tiered framework of security assurances, ensuring the comprehensive validation of computations. Individuals equipped with the appropriate hardware are empowered to take on the role of a *Worker* within the network, reaping rewards in return. Phala has implemented a range of protocols to guarantee that Workers consistently perform computations with unwavering fidelity and robust security. Phala Network combines tokenomic incentives, hardware-based security guarantees, and cryptographic validation on its blockchain. These elements work together to empower Phat Contracts that extend blockchain-level security to off-chain processes. Tokenomic incentives refer to the economic rewards tied to network participation, while hardware-rooted assurances involve secure hardware features. These combined features enhance security and facilitate off-chain computation, marking Phala a pioneer in the field of blockchain security and trustless computation.

Cosmos

Cosmos (`https://cosmos.network/`) stands out as a pioneering network in the blockchain landscape, aiming to solve some of the most pressing challenges faced by the industry: scalability, usability, and interoperability. By promoting a vision of creating an **Internet of Blockchains**, Cosmos facilitates an ecosystem where various blockchains can communicate and exchange information seamlessly, without sacrificing their sovereignty.

The Cosmos ecosystem, renowned for its innovative approach to blockchain interoperability and scalability, is witnessing a significant evolution with the integration of CosmWasm. **CosmWasm** is a pioneering smart contract platform that leverages the Rust programming language, offering a robust, secure, and flexible environment for developing **decentralized applications (dApps)**. This Rust-based framework is empowering new blockchains within the Cosmos network to achieve greater functionality and interoperability, marking a critical advancement in the ecosystem's capabilities.

CosmWasm stands out for its exceptional use of Rust, a programming language celebrated for its performance, reliability, and safety features. By utilizing Rust, CosmWasm ensures that smart contracts developed within the Cosmos ecosystem are not only efficient and fast, but also secure from common vulnerabilities. This is particularly important in the blockchain space, where security and trust are paramount. CosmWasm's integration into the Cosmos ecosystem allows developers to deploy complex smart contracts on various blockchains with ease, all while maintaining high levels of interoperability through the **Inter-Blockchain Communication (IBC)** protocol.

One of the core strengths of CosmWasm within the Cosmos ecosystem is its enhancement of interoperability among blockchains. Thanks to the IBC protocol and CosmWasm's flexible design, new and existing blockchains can seamlessly connect and share information, assets, and functionalities. This not only broadens the scope of possible applications but also simplifies the developer experience. Developers can now create dApps that leverage the strengths of multiple blockchains within the Cosmos network, using Rust to ensure their smart contracts are both powerful and secure.

The adoption of CosmWasm by new blockchains in the Cosmos ecosystem is rapidly growing thanks to its developer-friendly approach and the robust support provided by the Cosmos SDK. These new blockchains are utilizing CosmWasm to explore innovative use cases, from **decentralized finance (DeFi)** applications to **non-fungible tokens (NFTs)** and beyond. The flexibility of CosmWasm, combined with the Cosmos SDK, enables developers to tailor blockchain functionalities to specific needs, fostering a diverse and vibrant ecosystem of interconnected blockchains.

Cosmos represents a significant leap forward in the quest for a decentralized, interconnected blockchain ecosystem. With its innovative technologies and a strong focus on interoperability, scalability, and user-friendliness, Cosmos is paving the way for a new era of blockchain development. As it continues to grow and evolve, Cosmos is poised to become a cornerstone of the next generation of blockchain infrastructure, facilitating a more connected and efficient decentralized world.

Fuel

Fuel v1 (`https://www.fuel.network/`) originated as a **layer-2 (L2)** scalability solution tailored for a unified Ethereum infrastructure. It marked a significant milestone by becoming the inaugural optimistic rollup on the Ethereum *Mainnet*.

Despite the advent of L2 solutions that have contributed to a reduction in costs for accessing the Ethereum ecosystem, the overall increase in throughput has been rather restrained, both within optimistic and ZK approaches. **Zero-knowledge (ZK)** methods enhance privacy and scalability. For further details, reputable sources, such as Ethereum's official documentation, provide in-depth information on these techniques. During periods of heightened Ethereum network traffic, L2s have faltered in maintaining low transaction costs, frequently escalating to several dollars per transaction. The reason for this difficulty in maintaining low transaction costs is primarily due to the growing demand for Ethereum's limited computational resources during network congestion. As more users and applications compete for processing power, the cost of transactions tends to rise.

The evolution of **Layer-1 (L1)** blockchain architecture presents a transformative solution to the current challenges faced by L2 solutions, particularly in transaction costs. This shift moves away from tightly integrated consensus, data availability, and execution models, as observed in present-day Ethereum, towards the forthcoming modular framework. Here, execution is decoupled from data availability and consensus, exemplified by projects such as Eth2 and Celestia on the horizon. This decoupling fosters specialization at the foundational layer, resulting in a significant increase in bandwidth capacity, ultimately addressing the issue of escalating transaction costs in L2 systems.

Fuel is meticulously crafted to harness this enhanced bandwidth in ways unparalleled by any other scalability system. It stands as the swiftest execution layer within the modular blockchain stack, delivers supreme security, and is adaptable throughout. The term *flexible* carries substantial significance here, denoting the capacity for Ethereum-style interoperable Turing-complete smart contracts, extending beyond mere straightforward transfers. Referring to the ability to execute Ethereum-like interoperable Turing-complete smart contracts, this denotes a capacity that goes beyond simple token transfers. Turing-complete smart contracts are programs on a blockchain that can perform a wide range of computations and tasks, making them highly versatile and capable of complex operations beyond basic transactional functions. They enable the implementation of sophisticated logic and automation within blockchain networks.

Fuel's essence revolves around a modular execution layer, characterized as a verifiable computation system tailor-made for the modular blockchain stack. To put it more tangibly, Fuel encompasses a fraud- or validity-provable blockchain (or similar computation system) that capitalizes on a modular blockchain for ensuring data availability. It achieves this by leveraging a modular blockchain structure that prioritizes data availability. This approach ensures that transaction data is widely accessible and verifiable, reducing the risk of fraud or invalid transactions. In essence, Fuel's design combines elements of fraud-proof and validity-proof systems with a modular blockchain architecture to enhance the security and integrity of its transactions.

Fuel's specialization lies in optimizing execution efficiency to the fullest extent. This sets it apart from previously deployed rollup solutions that focused on the challenges inherent in a consolidated structure, such as constrained bandwidth. As the Ethereum ecosystem expands, ventures that fail to adapt will continue grappling with the repercussions of a compute-constrained design landscape. The moment for embracing modular execution has arrived.

With the various innovations it is spearheading, Fuel is the blockchain to watch out for.

Starcoin

Starcoin stands as a streamlined, promptly verified, and expandable application network, accompanied by secure native state cross-layer compatibility, establishing an infinitely scalable foundation for blockchain infrastructure.

Starcoin (`https://starcoin.org/en/`) ensures paramount security right from its inception through an elevated PoW consensus mechanism and a secure smart contract system executed in the Move programming language. Leveraging a stratified and adaptable approach to interoperability, developers of Starcoin equip decentralized financial networks with digital assets that empower all contributors in fostering a collaborative Web 3.0 ecosystem.

Utilizing Move, Starcoin has pioneered secure digital asset protocols encompassing FT and NFT. These protocols present formal verification tools, supplanting traditional contract auditing and bolstering contract and user asset security on the chain.

At Starcoin's Layer 1, an open and decentralized network of smart contracts beckons participation from all quarters, offering users ownership of their data through robust cryptographic principles and decentralized architecture. This layer remains fully attuned to real-time network congestion, dynamically tailoring network-wide configurations to optimize overall network utilization.

Meanwhile, Starcoin's Layer 2 manifests as an effective and scalable application network, facilitating secure and seamless state transfers, effortless data scaling, low latency, and exceptional operability.

Next, let's explore some innovative Rust projects that extend beyond traditional blockchain architectures, emphasizing their role as dynamic enablers for blockchain development across ecosystems.

Upcoming Rust Web3 projects

In this section, we delve into an intriguing realm that transcends traditional blockchain architecture. While the foundation of blockchain technology remains steadfast, a new horizon of innovation has emerged: projects that are not blockchains in themselves, but rather act as dynamic enablers for development across diverse blockchain ecosystems. These projects introduce a novel paradigm, unlocking unprecedented potential and fostering a symbiotic relationship with established blockchains.

Decentralization is crucial because it extends beyond individual blockchains. The projects we explore here showcase creativity and innovation, providing a vision of the future where collaboration, interoperability, and enhanced functionality push the boundaries of blockchain solutions.

These following projects basically enable quick and easy development on blockchains.

The Graph

The Graph Protocol (`https://thegraph.com/`) stands as a transformative pillar within the blockchain landscape, revolutionizing the way decentralized networks interact with data. At its core, The Graph empowers developers to seamlessly access, query, and extract insights from blockchain data, transcending the limitations of raw blockchain interactions. It does this by providing a decentralized indexing protocol and querying service that serves as the backbone of efficient data retrieval and analysis for dApps and services.

With its innovative architecture, The Graph serves as a bridge between developers and the vast universe of blockchain information, offering a user-friendly interface to create and manage subgraphs. These subgraphs act as tailored, modular data indexing layers, making it effortless for developers to extract specific data points and build immersive, data-rich dApps across a multitude of blockchain networks. Its architecture can be seen in the following figure:

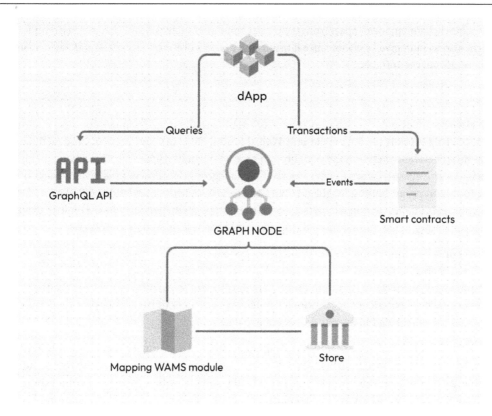

Figure 11.2 – Graph architecture

The Graph's impact reverberates through various sectors, including DeFi, NFTs, gaming, and beyond, where real-time data access and analytics play a pivotal role. Its decentralized nature ensures robustness and censorship resistance. Moreover, being built with Rust, being 100% open source, and growing rapidly, it stands as a top choice for smart contract engineers worldwide, enabling easier and quicker blockchain development.

Fe

Fe is more than just a project that helps with blockchain development, it is actually an entire programming language that enables smart contract development on the Ethereum blockchain, but the amazing part is that Fe is built with Rust.

Fe's design principles prioritize security by default, empowering developers to build with confidence. Its intuitive syntax and expressive capabilities facilitate seamless code composition, accelerating the development process while minimizing the risk of vulnerabilities. By integrating static analysis tools and leveraging the latest advancements in programming language research, Fe empowers developers to create robust and dependable smart contracts that mitigate common pitfalls.

If you're a blockchain engineer, apart from trying out languages such as Move that are based on Rust, you might also want to give Fe a shot since it brings the benefits of Rust – fast speed and performance – to Ethereum smart contracts.

Astar

Astar Network (`https://astar.network/`) stands as the premier smart contract platform in Japan, offering support for both EVM and **WebAssembly** (**Wasm**) environments and facilitating seamless interoperability between the two through a Cross-Virtual Machine. Embracing an inclusive approach, Astar Network caters to developers of all backgrounds, allowing them to harness their existing tools and programming languages. With the robust security foundation of Polkadot, Astar Network shines brilliantly within a dynamic and thriving ecosystem, playing a pivotal role in driving international corporate adoption and sparking consumer interest in Web3 technologies.

At the heart of Astar's innovation lies the **Build2Earn system** (`https://docs.astar.network/docs/learn/build2earn/`), ingeniously designed to foster network growth while rewarding contributors and builders. This system empowers developers to earn incentives for crafting and maintaining their decentralized applications, while users are incentivized to support their favored projects, thereby nurturing the overall ecosystem expansion.

Traditionally, blockchains face scalability challenges due to the security trade-offs inherent in decentralized consensus mechanisms. Astar Network is reshaping this landscape by transcending the limitations of independent blockchains and bridges, which often introduce vulnerabilities. Astar has achieved a revolutionary cross-chain synergy through its innovative utilization of both **Ethereum Virtual Machine** (**EVM**) and WebAssembly Virtual Machines. By seamlessly integrating these powerful computing engines, Astar combines the strengths of different blockchains, enabling them to work together cohesively. This approach allows developers to harness the best features from various blockchain platforms, ensuring the adaptability and future-proofing of their smart contract solutions. This paradigm shift in cross-chain functionality represents a significant advancement in blockchain technology.

Through the collaborative power of Polkadot's shared security and Astar's Cross-Virtual Machine, developers can now embark on a new era of smart contract creation. By fostering seamless cooperation and integration across various blockchains and applications, Astar Network pioneers the realization of unparalleled solutions that extend beyond replication.

Comit

Comit (`https://comit.network/`) is a really interesting project. While there are numerous existing blockchains such as Bitcoin and Ethereum, there are few projects facilitating communication between them. Without these connections, the Web3 landscape remains fragmented, with isolated blockchain communities. This hinders the realization of the Web3 vision, an interconnected internet of blockchains.

Comit aims to create this connection: it is a project that enables smooth cross-chain communication between multiple popular chains. This means moving assets easily between them, but Comit has achieved this without building another blockchain of its own. It's a solid project built in Rust, completely open source, and is one of the first of its kind. It is predicted that as the number of protocols increases, projects like these will definitely be required more and more and we will see more such projects spring up as more users adopt Web3.

Exonum

The **Exonum** platform (`https://exonum.com/index`), a robust open source blockchain framework, empowers enterprises and governments to confidently bring their blockchain initiatives to fruition. Leveraged across diverse industries by Bitfury's accomplished software engineering team, Exonum stands at the forefront of transaction speed and scalability in the industry. Its paramount security fortifies your blockchain endeavors.

Within the Exonum ecosystem, you can craft potent blockchain platforms enriched with smart contracts. This is coupled with the ability to anchor to public blockchains and more. Specifically designed for seamless integration with public blockchains like Bitcoin's, Exonum's anchoring process captures a *snapshot* of a system state, obviating the need for unconditional trust in blockchain administrators. This approach maintains data confidentiality while harnessing the robust security of public blockchains. Consequently, any modifications to your Exonum blockchain, should a malicious entity compromise the majority of nodes, are documented and detected swiftly.

Augmenting its security measures, the Exonum platform is meticulously developed using Rust, a highly secure programming language. Rust ensures comprehensive execution safety and predictable resource utilization. Moreover, Exonum provides you the capability to restrict data visibility in your blockchain, safeguarding user privacy without compromising security. Additionally, Exonum employs a custom **Byzantine consensus algorithm**, bolstering data integrity without necessitating resource-intensive *mining* computations. This resilience remains intact even when nodes experience malfunctions or susceptibility to manipulation.

To facilitate streamlined audits of your Exonum blockchain, the platform seamlessly integrates into various customer applications, enhancing transparency and ease of oversight.

Now that we have learnt quite a bit about blockchain projects and other projects that help with blockchain development that use Rust, let's talk about how the Rust community is gearing up for the blockchain revolution, and then about jobs in the Web3 space.

The Rust community

Rust solves a lot of problems that are present with other programming languages, and we have already learned about the revolutionary features of Rust. Rust is quite disruptive. Recently, Windows rewrote the entire Kernel of Windows with just 180,000 lines of Rust code – that's much less code to power the entire Windows Kernel. This just goes to show how robust Rust is as a technology and how disruptive it will be in the software, operating system, and web worlds.

Since this book is geared towards *Web3*, let's talk about what the future looks like.

Even though Rust is highly effective and loved by a lot of programmers, the traction for Rust has been a bit on the lower side; that's because it's not a language like JavaScript or Python. While Java and Python are easy to pick up because of the lack of complex memory management or memory borrowing, Rust, on the other hand, has a need for complex memory management with a lot of features that may require learning and understanding programming at a deeper level, and not everyone is interested in doing this.

Since the number of Web3 projects that use Rust is growing at an unprecedented rate, as we have just seen in this chapter, the demand for Web3 engineers who know Rust will keep on growing. What's even happening in the industry is that there are engineers that want to work on these projects and the only reason they're learning Rust is because they want to work on the blockchain side of things. This is a huge shift from earlier, where developers knew Rust and only later explored blockchain engineering.

So, we're going to see more developers enter into the Rust ecosystem just to be able to work on blockchains, and that's a great thing since it would mean significant growth for Rust and a rich community of blockchain developers within the Rust community.

This goes on to bolster our earlier observation that the future for Rust and blockchains is really bright, and this might just be the right time to start a career in blockchain and Rust (`https://www.rust-lang.org/community`).

Now, let's delve into the various job roles and career opportunities within the Web3 space, where your Rust skills can shine.

Jobs in the Web3 space

If you're reading this book, there's a high chance you might be really interested in starting a career as a blockchain developer or engineer specializing in Rust. Since there are so many Web3 projects that leverage Rust, as we have seen, it is only logical that in the near future there will be significant demand for Rust blockchain engineers.

In this section, we will go over a few details that could be helpful to you in starting a career in Rust blockchain engineering.

Popular job roles

The most popular job roles are without a doubt related to engineering in blockchain, but there are also product managers, program managers, experience designers, and token managers or designers peripheral to engineering roles. Generally, blockchain is quite a technical field and even the marketing and community managers are also quite technical. Apart from these roles, there are other new types of roles that have played a great role in the Web3 world; one such role is that of the Dev-Rel, or developer relations manager.

Let's go through each of these job roles one by one.

Engineering roles are usually of two types: *smart contract engineers* and *protocol engineers*. The protocol engineers work on the core blockchain product, which requires a deep level of backend engineering expertise. Smart contract engineers focus on building dApps; they work with smart contract programming languages such as Move, Fe, and Solidity.

There are also *security engineers*, which can be of many types. Some are focused only on smart contract security and this is why they're called smart contract security engineers. Web3 security isn't just restricted to smart contract auditing but goes way beyond, and this is why *Web3 security engineers* need to know about Pentesting tools, attack simulation, threat monitoring, etc. as well.

Product managers are responsible for managing the product roadmap and for the communication and alignment between various teams since the product roadmap for blockchains ends up being highly technical. It helps if the product manager already has technical experience. *Program managers* basically manage the technical work and details for the project and are responsible for ensuring everything is happening as intended based on the product roadmap. Their job is all about converting business requirements into technical requirements for the tech teams. Needless to say, they need to be more technically inclined than the product manager.

Then there are *technical architects*, or more specifically *blockchain architects*, who are involved in the planning and design of how the Blockchain will function. They come up with a blueprint, and that's what the entire engineering team follows. The **engineering manager** (**EM**) is responsible for ensuring timely deliveries. The EM manages the engineers based on the blueprint from the architect and the timelines set by the program manager, and this is how the entire team works together in delivering the blockchain project.

To interact with the blockchain, frontend dashboards are also required, and this is why there are *UX designers* andfrontend *engineers* in the project, but their work is minimal in the blockchain world as more and more projects are relying on no-code tools to accomplish this.

Token design and management is a science because you need to think about utility, growth, and value exchange with tokens. From everything related to distribution to vesting to governance has to be planned by the *token managers*. Their work is complex and requires expertise.

Finally, let's talk about *Dev-Rel*. Now, *marketing and community managers* have been important for digital products since the beginning, and community makes a lot of sense for Web3 projects since many of the projects are extremely technical and require dedicated communities that understand what the product does. In many cases, the community contributes to the development and growth of the product or platform.

Dev-Rel is just an extension of this and is a common job role in the Web3 world. Dev-Rel managers are responsible for ensuring that the latest technical docs are available, the engineering communities have all that they require to build with, and there's enough education about the product on the market that developers are able to adopt the product. The target audience for most of the products is engineers, and success is measured based on the number of projects built by engineers for the technology.

Now that we have clarity on the many job roles that are available in the blockchain and Web3 world, it's time to learn about how you can find these jobs.

How to find Web3 jobs

Once you have a good command of Rust and some blockchain engineering, it'll be easy to find jobs. Let's talk about a few platforms that make job searching really simple for blockchain developers:

- *CryptoJobs*: A well-known platform that focuses specifically on job listings in the cryptocurrency and blockchain space. You can filter jobs by category, such as development, engineering, design, and more. See more at `https://www.cryptojobs.com/`.

- *AngelList*: A platform that features a section dedicated to blockchain and cryptocurrency job listings. You can find opportunities in various roles, including engineering, development, marketing, and management. See more at `https://www.angellist.com/`.

- *LinkedIn*: A versatile platform where you can network with professionals in the blockchain industry and discover job opportunities posted by companies. Follow relevant hashtags and join blockchain-related groups to stay updated. See more at `https://www.linkedin.com/`.

- *Indeed*: A widely used job search engine that includes a category for blockchain-related positions. You can search for roles such as blockchain engineer, Rust developer, smart contract developer, and more. See more at `https://www.indeed.com`.

- *Dice*: A platform with a focus on tech jobs. You can often find blockchain and Rust-related positions listed here. It's especially useful for finding technology-specific roles. See more at `https://www.dice.com/`.

- *GitHub Jobs*: A platform where companies post developer and engineering roles, including those related to Rust and blockchain. You can filter by location and skills to find relevant opportunities. See more at `https://github.com/about/careers`.

- *Ethlance*: A decentralized job marketplace built on the Ethereum blockchain. It's designed for freelancers and allows you to browse or list jobs related to blockchain and Rust development. See more at `https://www.etherlance.io/`.

- *Blockchain Developer Jobs*: A dedicated website that provides a curated list of blockchain development jobs from various sources, making it easy to find opportunities in the blockchain engineering field.

- *Crypto Careers*: A website that aggregates job listings from different sources, helping you discover blockchain and Rust-related roles from various platforms. See more at `https://www.crypto-careers.com/`.

- *X (previously Twitter)*: A site where you can follow blockchain influencers, developers, and companies. They often share job openings and opportunities within the industry. Use hashtags such as `#BlockchainJobs` or `#RustJobs` to find relevant posts. See more at `www.X.com`.

- *Reddit*: A forum social network with several subreddits dedicated to job listings in the blockchain and Rust development fields. Subreddits such as *r/BlockchainJobs* and *r/RustJobs* can be helpful resources.

- *Company websites*: Visit the career sections of blockchain companies' websites directly. Many companies in the blockchain space, including those focusing on Rust development, list job openings on their official sites. You could also visit `https://web3.career/` to explore career opportunities.

Building a career

Getting a job is very different from building a strong career in blockchain engineering. Ideally, the play should always be long-term, since this builds credibility, so let's talk about some ways, methods, and means that you can follow to build a fruitful career in Web3 engineering:

- **Relevant skills**: Start by building a strong foundation in blockchain-related skills, such as smart contract development, consensus algorithms, cryptography, decentralized applications, and blockchain platforms such as Ethereum, Binance Smart Chain, or Polkadot. Proficiency in programming languages such as Solidity, Rust, or JavaScript can be particularly valuable.

- **Education and training**: Consider formal education, online courses, workshops, and certifications that specialize in blockchain technology. Platforms such as Coursera, Udemy, and edX offer a variety of blockchain-related courses. Earning certificates or degrees in computer science, software engineering, or related fields can also enhance your qualifications.

- **Networking**: Attend blockchain meetups, conferences, and webinars in your area or online. Engaging with professionals and enthusiasts in the blockchain space can provide valuable insights, job leads, and opportunities to showcase your skills.

- **Online platforms**: Utilize online job platforms and websites that focus on blockchain and cryptocurrency jobs, such as CryptoJobs, AngelList, LinkedIn, Indeed, and specialized blockchain forums. Set up job alerts to receive notifications about relevant openings.

- **Open source contributions**: Contribute to open source blockchain projects on platforms such as GitHub. This not only showcases your skills but also allows you to collaborate with experienced developers and establish a presence in the community.

- **Personal projects**: Create your own blockchain projects, smart contracts, or dApps to demonstrate your capabilities. A strong portfolio of practical examples can set you apart from other candidates.

- **LinkedIn profile**: Optimize your LinkedIn profile to highlight your blockchain skills, projects, and experience. Connect with professionals in the blockchain industry to expand your network.

- **Job boards and forums**: Explore blockchain-focused job boards and forums where companies post openings. Engage in discussions, ask questions, and keep an eye out for job postings.

- **Internships and entry-level positions**: Consider starting with internships or entry-level positions to gain hands-on experience and learn from seasoned professionals in the field.

- **Tailored applications**: Customize your application materials (resume, cover letter, portfolio) for each job you apply to. Highlight relevant skills, projects, and experiences that align with the specific role.

- **Soft skills**: In addition to technical skills, emphasize soft skills such as problem-solving, teamwork, communication, and adaptability. These qualities are highly valued in the fast-paced and collaborative blockchain industry.

- **Continuous learning**: Stay updated with the latest developments in the blockchain field, as technology evolves rapidly. Engage in continuous learning and adapt your skills to emerging trends.

Going beyond this book

In closing, this journey through the realm of Rust-powered blockchain development has equipped you with a robust toolkit of skills and insights, paving the way for your continued exploration of the Web3 landscape. As you reflect on the knowledge gained within these pages, you stand at the threshold of limitless possibilities, poised to embark on an exciting voyage of building, innovating, and transforming the decentralized world.

With a solid foundation in Rust and blockchain intricacies, your potential to contribute to the evolving blockchain ecosystem is boundless. Take the lessons learned from building your own blockchain and extending your expertise to the Rust-based ecosystems of Solana, Near, Polkadot, and beyond. Seize the opportunity to craft innovative solutions, bridging the gaps and shaping the future of decentralized technologies.

To extend your journey beyond these chapters, consider the following pathways:

- **Forge your own creations**: Utilize your enhanced skills to pioneer innovative projects, exploring the possibilities of Rust and blockchain synergy. Whether crafting pioneering decentralized applications, advanced smart contracts, or inventive cross-chain interoperability solutions, the opportunity to innovate is boundless. Dive into projects such as decentralized exchanges, NFT marketplaces, privacy-centric cryptocurrencies, and secure voting systems. Unleash creativity by merging strong coding expertise with the game-changing potential of blockchain technology.

- **Join the collaborative realm of Web3**: Engage with the vibrant Web3 community by contributing to open source projects, participating in hackathons, and attending blockchain meetups and conferences. Collaborating with fellow developers will not only accelerate your learning but also broaden your horizons through shared experiences. The collaborative platforms can include decentralized version control systems such as Git, collaborative coding environments such as GitHub, and decentralized project management tools. See more at `https://github.com/topics/blockchain-projects`.

- **Pursue a fulfilling Web3 career**: As the Web3 ecosystem continues to expand, opportunities for impactful careers in blockchain development are abundant. Consider roles such as blockchain engineer, smart contract developer, protocol designer, or decentralized application architect, and let your expertise propel you to the forefront of this transformative industry.

- **Stay current and evolve**: The blockchain landscape is ever-evolving, with new technologies, standards, and paradigms continuously emerging. Commit to staying up to date with the latest advancements, experimenting with cutting-edge tools, and embracing the dynamic nature of this realm. Consider exploring resources such as blockchain-focused forums (e.g., r/blockchain, Ethereum Stack Exchange), Rust programming communities (e.g., r/Rust, Rust Discord channels), blogs (e.g., Rust official blog, Ethereum Foundation blog), and reputable tech news platforms (e.g., Coindesk, CoinTelegraph). Additionally, following influential figures and thought leaders in the blockchain and Rust space on social media platforms, such as Twitter, can provide valuable insights into emerging trends and technologies. Remember to also keep an eye on official documentation and announcements from Rust and prominent blockchain projects.

Summary

In this chapter, we explored Rust's future in the blockchain world and the career prospects it offers to Rust engineers and blockchain develop ers. We examined Rust's role in various blockchains and Web3 technologies, delved into the growing Rust community's involvement in blockchain, and explored job opportunities in the Web3 space. Lastly, we discussed additional blockchain projects and how to extend your blockchain develoment journey.

With that, we come to the end of this book. As you step into the next phase of your journey, remember that the knowledge you've acquired here is merely the foundation. The future beckons with challenges and opportunities that await your unique imprint. Your dedication to mastering Rust and blockchain development has positioned you as an agent of change in the Web3 era. Embrace the evolution, champion innovation, and continue to explore the boundless frontiers of this decentralized world.

May your path be illuminated by the spirit of innovation and guided by the ever-present promise of the blockchain horizon. Onward, dear reader, to a future that you will help to shape and redefine.

Index

packtpub.com

Subscribe to our online digital library for full access to over 7,000 books and videos, as well as industry leading tools to help you plan your personal development and advance your career. For more information, please visit our website.

Why subscribe?

- Spend less time learning and more time coding with practical eBooks and Videos from over 4,000 industry professionals

- Improve your learning with Skill Plans built especially for you

- Get a free eBook or video every month

- Fully searchable for easy access to vital information

- Copy and paste, print, and bookmark content

Did you know that Packt offers eBook versions of every book published, with PDF and ePub files available? You can upgrade to the eBook version at packtpub.com and as a print book customer, you are entitled to a discount on the eBook copy. Get in touch with us at customercare@packtpub.com for more details.

At www.packtpub.com, you can also read a collection of free technical articles, sign up for a range of free newsletters, and receive exclusive discounts and offers on Packt books and eBooks.

Other Books You May Enjoy

If you enjoyed this book, you may be interested in these other books by Packt:

Blockchain Development for Finance Projects

Ishan Roy

ISBN: 978-1-83882-909-4

- Design and implement blockchain solutions in a BFSI organization
- Explore common architectures and implementation models for enterprise blockchain
- Design blockchain wallets for multi-purpose applications using Ethereum
- Build secure and fast decentralized trading ecosystems with Blockchain
- Implement smart contracts to build secure process workflows in Ethereum and Hyperledger Fabric
- Use the Stellar platform to build KYC and AML-compliant remittance workflows
- Map complex business workflows and automate backend processes in a blockchain architecture

Securing Blockchain Networks like Ethereum and Hyperledger Fabric

Alessandro Parisi

ISBN: 978-1-83864-648-6

- Understand blockchain consensus algorithms and security assumptions
- Design secure distributed applications and smart contracts
- Understand how blockchains manage transactions and help to protect wallets and private keys
- Prevent potential security threats that can affect distributed ledger technologies (DLTs) and blockchains
- Use pentesting tools for assessing potential flaws in Dapps and smart contracts
- Assess privacy compliance issues and manage sensitive data with blockchain

Packt is searching for authors like you

If you're interested in becoming an author for Packt, please visit `authors.packtpub.com` and apply today. We have worked with thousands of developers and tech professionals, just like you, to help them share their insight with the global tech community. You can make a general application, apply for a specific hot topic that we are recruiting an author for, or submit your own idea.

Share Your Thoughts

Now you've finished *Rust for Blockchain Application Development*, we'd love to hear your thoughts! Scan the QR code below to go straight to the Amazon review page for this book and share your feedback or leave a review on the site that you purchased it from.

`https://packt.link/r/1837634645`

Your review is important to us and the tech community and will help us make sure we're delivering excellent quality content.

Download a free PDF copy of this book

Thanks for purchasing this book!

Do you like to read on the go but are unable to carry your print books everywhere?

Is your eBook purchase not compatible with the device of your choice?

Don't worry, now with every Packt book you get a DRM-free PDF version of that book at no cost.

Read anywhere, any place, on any device. Search, copy, and paste code from your favorite technical books directly into your application.

The perks don't stop there, you can get exclusive access to discounts, newsletters, and great free content in your inbox daily

Follow these simple steps to get the benefits:

1. Scan the QR code or visit the link below

https://packt.link/free-ebook/9781837634644

2. Submit your proof of purchase
3. That's it! We'll send your free PDF and other benefits to your email directly

www.ingramcontent.com/pod-product-compliance
Lightning Source LLC
Chambersburg PA
CBHW080610060326
40690CB00021B/4638